Descriptive Chemistry of the Elements

Table of Atomic Masses and Numbers

Based on the 1991 Report of the Commission on Atomic Weights and Isotopic Abundances of the International Union of Pure and Applied Chemistry and for the elements as they exist naturally on earth. Scaled to the relative atomic mass of carbon-12. The estimated uncertainties in values, between ±1 and ±9 units in the last digit of an atomic mass, are in parentheses after the atomic mass. (From *Journal of Physical and Chemical Reference Data,* Vol. 22(1993), pp. 1571–1584. © 1993 IUPAC)

Element	Symbol	Atomic Number	Atomic Mass	
Actinium	Ac	89	227.0278	(L)
Aluminum	Al	13	26.981539(5)	
Americium	Am	95	243.0614	(L)
Antimony	Sb	51	121.757(3)	
Argon	Ar	18	39.948(1)	(g, r)
Arsenic	As	33	74.92159(2)	
Astatine	At	85	209.9871	(L)
Barium	Ba	56	137.327(7)	
Berkelium	Bk	97	247.0703	(L)
Beryllium	Be	4	9.012182(3)	
Bismuth	Bi	83	208.98037(3)	
Boron	B	5	10.811(5)	(g, m, r)
Bromine	Br	35	79.904(1)	
Cadmium	Cd	48	112.411(8)	(g)
Calcium	Ca	20	40.078(4)	(g)
Californium	Cf	98	251.0796	(L)
Carbon	C	6	12.011(1)	(g, r)
Cerium	Ce	58	140.115(4)	(g)
Cesium	Cs	55	132.90543(5)	
Chlorine	Cl	17	35.4527(9)	(m)
Chromium	Cr	24	51.9961(6)	
Cobalt	Co	27	58.93320(1)	
Copper	Cu	29	63.546(3)	(r)
Curium	Cm	96	247.07003	(L)
Dysprosium	Dy	66	162.50(3)	(g)
Einsteinium	Es	99	252.083	(L)
Erbium	Er	68	167.26(3)	(g)
Europium	Eu	63	151.965(9)	(g)
Fermium	Fm	100	257.0951	(L)
Fluorine	F	9	18.9984032(9)	
Francium	Fr	87	223.0197	(L)
Gadolinium	Gd	64	157.25(3)	(g)
Gallium	Ga	31	69.723(4)	
Germanium	Ge	32	72.61(2)	
Gold	Au	79	196.96654(3)	
Hafnium	Hf	72	178.49(2)	
Hahnium	Ha	105	262.114	(L, n)
Hassium	Hs	108	265	(n)
Helium	He	2	4.002602(2)	(g, r)
Holmium	Ho	67	164.93032(3)	
Hydrogen	H	1	1.00794(7)	(g, m, r)
Indium	In	49	114.818(3)	
Iodine	I	53	126.90447(3)	
Iridium	Ir	77	192.22(3)	
Iron	Fe	26	55.847(3)	
Krypton	Kr	36	83.80(1)	(g, m)
Lanthanum	La	57	138.9055(2)	(g)
Lawrencium	Lr	103	262.11	(L)
Lead	Pb	82	207.2(1)	(g, r)
Lithium	Li	3	6.941(2)	(g, m, r)
Lutetium	Lu	71	174.967(1)	(g)
Magnesium	Mg	12	24.3050(6)	
Manganese	Mn	25	54.93805(1)	
Meitnerium	Mt	109	266	(n)
Mendelevium	Md	101	258.10	(L)
Mercury	Hg	80	200.59(2)	
Molybdenum	Mo	42	95.94(1)	(g)
Neodymium	Nd	60	144.24(3)	(g)
Neon	Ne	10	20.1797(6)	(g, m)
Neptunium	Np	93	237.0482	(L)
Nickel	Ni	28	58.6934(2)	
Nielsbohrium	Ns	107	262.12	(L, n)
Niobium	Nb	41	92.90638(2)	
Nitrogen	N	7	14.00674(7)	(g, r)
Nobelium	No	102	259.1009	(L)
Osmium	Os	76	190.23(3)	(g)
Oxygen	O	8	15.9994(3)	(g, r)
Palladium	Pd	46	106.42(1)	(g)
Phosphorus	P	15	30.973762(4)	
Platinum	Pt	78	195.08(3)	
Plutonium	Pu	94	244.0642	(L)
Polonium	Po	84	208.9824	(L)
Potassium	K	19	39.0983(1)	(g)
Praseodymium	Pr	59	140.90765(3)	
Promethium	Pm	61	144.9127	(L)
Protactinium	Pa	91	231.03588(2)	
Radium	Ra	88	226.0254	(L)
Radon	Rn	86	222.0176	(L)
Rhenium	Re	75	186.207(1)	
Rhodium	Rh	45	102.90550(3)	
Rubidium	Rb	37	85.4678(3)	(g)
Ruthenium	Ru	44	101.07(2)	(g)
Rutherfordium	Rf	104	261.11	(L, n)
Samarium	Sm	62	150.36(3)	(g)
Scandium	Sc	21	44.955910(9)	
Seaborgium	Sg	106	263.118	(L, n)
Selenium	Se	34	78.96(3)	
Silicon	Si	14	28.0855(3)	(r)
Silver	Ag	47	107.8682(2)	(g)
Sodium	Na	11	22.989768(6)	
Strontium	Sr	38	87.62(1)	(g, r)
Sulfur	S	16	32.066(6)	(g, r)
Tantalum	Ta	73	180.9479(1)	
Technetium	Tc	43	98.9072	(L)
Tellurium	Te	52	127.60(3)	(g)
Terbium	Tb	65	158.92534(3)	
Thallium	Tl	81	204.3833(2)	
Thorium	Th	90	232.0381(1)	(g)
Thulium	Tm	69	168.93421(3)	
Tin	Sn	50	18.710(7)	(g)
Titanium	Ti	22	47.88(3)	
Tungsten	W	74	183.84(1)	
Ununnilium	Uun	110	269	L
Unununium	Uuu	111	272	L
Uranium	U	92	238.0289(1)	(g, m)
Vanadium	V	23	50.9415(1)	
Xenon	Xe	54	131.29(2)	(g, m)
Ytterbium	Yb	70	173.04(3)	(g)
Yttrium	Y	39	88.90585(2)	
Zinc	Zn	30	65.39(2)	
Zirconium	Zr	40	91.224(2)	(g)

(g) Geologically exceptional specimens of this element are known that have different isotopic compositions. For such samples, the atomic mass given here may not apply as precisely as indicated.

(L) This atomic mass is the relative mass of the isotope of longest half-life. The element has no stable isotopes.

(m) Modified isotopic compositions can occur in commercially available materials that have been processed in undisclosed ways, and the atomic mass given here might be quite different for such samples.

(n) Name and symbol approved in 1995 for use in the United States by the Nomenclature Committee of the American Chemical Society.

(r) Ranges in isotopic compositions of normal samples obtained on earth do not permit a more precise atomic mass for this element, but the tabulated value should apply to any normal sample of the element.

Descriptive Chemistry of the Elements

James E. Brady

St. John's University, New York

John R. Holum

Augsburg College (Emeritus), Minnesota

John Wiley & Sons, Inc.

New York / Chichester / Brisbane / Toronto / Singapore

ACQUISITIONS EDITOR Nedah Rose
DEVELOPMENTAL EDITOR Kathleen Dolan
MARKETING MANAGER Catherine Faduska
SENIOR PRODUCTION EDITOR Suzanne Magida
DESIGNER Kevin Murphy
MANUFACTURING MANAGER Mark Cirillo
PHOTO EDITOR Lisa Passmore
ILLUSTRATION EDITOR Sigmund Malinowski

This book was set in 10/12 New Baskerville by Progressive Information Technologies and printed and bound by Von Hoffman Press. The cover was printed by Phoenix Color.

Electronic Illustrations were provided by Fine Line.

Recognizing the importance of preserving what has been written, it is a policy of John Wiley & Sons, Inc. to have books of enduring value published in the United States printed on acid-free paper, and we exert our best efforts to that end.

The paper on this book was manufactured by a mill whose forest management programs include sustained yield harvesting of its timberlands. Sustained yield harvesting principles ensure that the number of trees cut each year does not exceed the amount of new growth.

James E. Brady and John R. Holum
Descriptive Chemistry of the Elements
ISBN 0-471-13557-7

Printed in the United States of America

10 9 8 7 6 5 4 3 2 1

Preface

This Supplement was written to accompany the text *Chemistry: The Study of Matter and Its Changes.* Its goal is to provide a discussion of the descriptive chemistry of the elements at a level appropriate for the first-year general chemistry course for science students.

Today, many teachers find that most of their students do not need the traditional in-depth treatment of the descriptive chemistry of the elements. Time is spent instead discussing chemical principles, with descriptive chemistry serving to illustrate principles when appropriate. By separating the chapters concentrating on descriptive chemistry from the main text, our aim was to satisfy this audience. However, we also realize that there is a significant number of teachers who believe that descriptive chemistry plays an important role in the education of a chemistry student, and who do not wish to abandon teaching these topics. This book recognizes the importance of this viewpoint and presents an organized discussion of the properties and reactions of the chemical elements and their compounds.

Besides being used as a supplement to the textbook mentioned above, some teachers may wish to use this book as a supplement in a sophomore inorganic chemistry course. In this context, it would enable chemistry students who have not been exposed to descriptive inorganic chemistry to study these topics without the rigors of physical chemistry as background.

ORGANIZATION

We have elected to discuss the chemistry of the nonmetallic elements first, followed by metals. In doing this, we have not attempted to cover the chemistry of every element, but instead have focused our attention on those that are most common and important. We begin, in Chapter 1, with the elements hydrogen, oxygen, nitrogen, and carbon—the four most common elements in living systems. We conclude our discussion of the nonmetals in Chapter 2 with sulfur, phosphorus, the halogens, and the noble gases (the latter to emphasize that these elements are not inert).

In the previous edition, our discussion of the metallic elements was divided between two chapters. We have condensed this to a single chapter after placing the treatment of metal complexes in a separate chapter in the main text. In this final supplement chapter, we examine the representative elements in some detail. We have not attempted to present the chemistry of all the transition metals, choosing instead to select the most common of them.

James E. Brady
John R. Holum

Contents

Descriptive Chemistry of the Elements

This imperial crown of the Russian Tsars displays perhaps the most admired gem in the world, the diamond, a form of carbon, one of the elements studied in this chapter.

Chapter 1

Simple Molecules and Ions of Nonmetals: Part I

1.1 THE PREVALENCE OF NONMETALLIC ELEMENTS

Whether we journey to the outer reaches of space or plunge to the molecular level of life, a handful of *nonmetallic* elements of low atomic mass make up most of the universe. In this chapter and the next we will study their most familiar chemical and physical properties. They include the two lightest elements, hydrogen and helium. Together, these account for 99.9% of all of the atoms in existence! Hydrogen and helium so dominate the composition of the universe that all of the other elements seem like impurities. How has this probably come about?

An estimate like this is based on the frequencies and relative intensities of the lines in the spectra of radiation from stars as well as the analyses of meteorites.

Big Bang Theory

According to the theory most widely accepted today, the formation of all stars, suns, planets, and other cosmic bodies can be traced to one original body. To explain the present rate of expansion of the universe, its estimated mass, and the kind of background radiation currently observed, this initial body must have had a density of about 10^{96} g cm^{-3} and a temperature of about 10^{32} K. It exploded—hence, the theory is named the "big bang" theory—and in just one second its temperature fell to about 10^{10} K as the dense material expanded. The products of the explosion consisted of about equal numbers of protons and neutrons as well as electrons, and for about 8 minutes the system operated like a huge nuclear fusion furnace (see Section 20.8).[1] By a number of complex nuclear transformations, all of the elements eventually formed from these elementary particles. In various kinds of stars, these processes continue today, but, overall, hydrogen and helium still are the dominant elements.

At the sun's center, $T = 1.5 \times 10^7$ K; the density is 1.5×10^2 g cm^{-3}.

[1] Unless noted otherwise, cross-references to Sections, Special Topics, and *Chemicals in Use* refer to the accompanying text, *Chemistry: The Study of Matter and Its Changes,* 2nd edition (1996), by Brady and Holum, not to this supplement. Some material from text chapters 15 to 19 is assumed for this supplement.

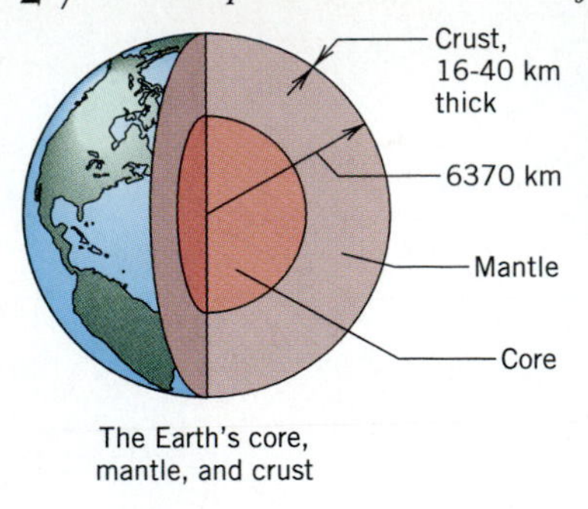

The Earth's core, mantle, and crust

The Earth's Elements

Atoms of the higher elements are largely concentrated in various bodies of the universe, yet the atoms of one nonmetal, oxygen, still dominate. Oxygen atoms are the most common in the Earth, contributing 50% of the Earth's atoms, overall, and 61% of the atoms of the Earth's crust. The Earth's atmosphere is 78% nitrogen and 21% oxygen (by volume) with carbon dioxide, argon, and other noble gases supplying nearly all of the rest. Nonmetals—hydrogen, oxygen, nitrogen, carbon, sulfur, phosphorus, and chlorine—also dominate the human body, making up over 87% of its mass. This chapter is devoted to the first four.

1.2 HYDROGEN

Hydrogen is a colorless, odorless, tasteless gas existing under normal conditions as diatomic molecules, H_2. It boils at 20 K and freezes at 14 K.

Preparation of Hydrogen

Hydrocarbons are compounds of carbon and hydrogen that occur in natural gas and oil. Propane is just one example.

Industrial Preparation Most hydrogen is made industrially by stripping it from molecules of natural gas or oil refinery hydrocarbons. Using propane, C_3H_8, as an example, a mixture of the hydrocarbon and high-temperature steam is passed over a catalyst at about 900 °C where the following reaction occurs irreversibly.

$$C_3H_8 + 3H_2O \xrightarrow[\text{catalyst}]{900\ ^\circ\text{C}} 3CO + 7H_2$$

The carbon monoxide and hydrogen are then separated by a combination of chemical and physical methods.

In labs where hydrogen is frequently used, it is obtained from cylinders of compressed hydrogen.

Small-Scale Preparations of Hydrogen Hydrogen can be made in the lab by the reaction of a metal such as zinc with hydrochloric acid (see Figure 1.1).

FIGURE 1.1

(*a*) Hydrogen gas bubbles vigorously from a test tube in which zinc metal reacts with hydrochloric acid. (*b*) In an apparatus such as this, hydrogen gas can be collected in the lab by the reaction of zinc with hydrochloric acid.

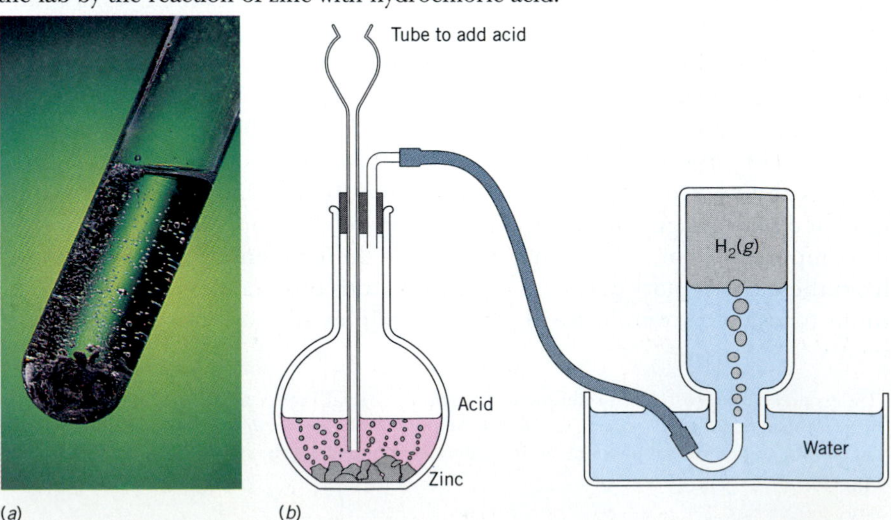

In this oxidation–reduction reaction, zinc is oxidized and the hydrogen ion reduced.

$$Zn(s) + 2HCl(aq) \longrightarrow H_2(g) + ZnCl_2(aq)$$

Small quantities of hydrogen are also made by the reaction of calcium metal with water. Calcium, being a stronger reducing agent than zinc, is able to reduce water.

Henry Cavendish (1731–1810), an English chemist credited as the first to recognize hydrogen as a substance, made it this way.

$$Ca(s) + 2H_2O \longrightarrow Ca(OH)_2(s) + H_2(g)$$

Isotopes of Hydrogen

Hydrogen is the only element whose isotopes have their own names. The most abundant is commonly called *hydrogen,* but *protium* is used in special situations. Protium accounts for over 99% of all hydrogen (see Table 1.1).

The Deuterium Isotope Effect Although an element's isotopes generally give identical *kinds* of chemical reactions, the *rates* of these reactions can differ. Both protium and deuterium, for example, react with chlorine; H_2 reacts to give HCl, and D_2 reacts to give DCl. At 0 °C, however, the reaction of H_2 with chlorine is 13.4 times as rapid as that of D_2.

A difference in rates shown by different isotopes in the same kind of reaction is called an **isotope effect,** and deuterium's generally slower reactivity than protium is called the **deuterium isotope effect.** It is the most pronounced isotope effect of all that have been studied, and it originates in relative bond strengths. Covalent bonds to D are generally stronger than those to H, so the energy of activation for a rate-limiting elementary process (see Section 14.8) is greater for breaking a bond to D than to H.

As we have learned in Chapter 14, a larger energy of activation means a slower rate.

Chemical Properties of Hydrogen

We will continue the study of hydrogen as if it consisted of just the common isotope. The binary compounds of hydrogen with a second element are called **hydrides,** and their properties depend on whether the second element is a metal or a nonmetal. Some binary hydrides are ionic compounds, others are covalent, and those with certain transition metals are in a class by themselves (see the figure in the margin).

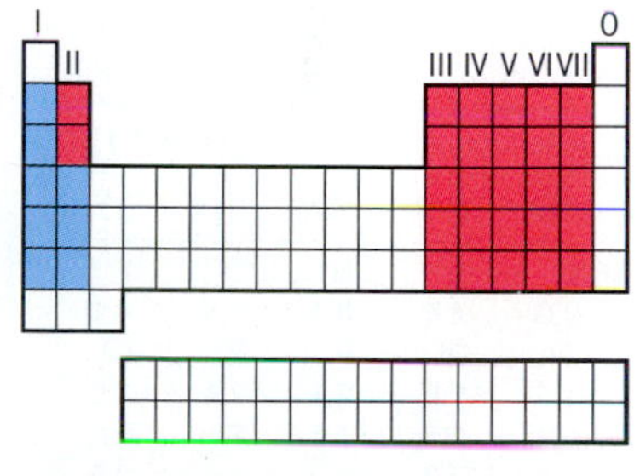

Among the representative elements:
Form ionic hydrides
Form covalent hydrides

Ionic Hydrides The ionic hydrides are made by a direct combination of hydrogen with the metal at high temperatures, for example,

$$Ca(s) + H_2(g) \xrightarrow{300-400\ °C} CaH_2(s)$$

Hydrogen occurs as the *hydride ion,* H^-, in the ionic hydrides, which conduct electricity at or just below their melting points. Hydrogen gas forms at the

TABLE 1.1 Isotopes of Hydrogen

Isotope Name	Atomic Symbol	Mass Number	Atom Percent	Melting Point (K)	Boiling Point (K)
Protium	H	1	99.985	13.957	20.39
Deuterium	D	2	0.015	18.73	23.67
Tritium	T	3	Trace[a]	20.62	25.04

[a] Radioactive; half-life, 12.35 years; total atmospheric content, about 11 g.

Ionic hydrides are dissolved in molten alkali halides for electrolysis.

anode, which is evidence that hydride ions are present. The most likely anode reaction is the oxidation of H^-.

$$2H^- \longrightarrow H_2 + 2e^-$$

The H^- ion is a powerful electron donor, so it can exist in ionic compounds only with cations that are the least able to accept electrons. Thus, only the metals with the lowest ionization energies form *ionic* hydrides, generally those from Groups IA and IIA, except beryllium and magnesium in Group IIA, which form hydrides that are more covalent than ionic.

Bubbles of H_2 evolve as CaH_2 reacts with water.

Because the hydride ion is both an exceptionally powerful Brønsted base and a strong reducing agent, the ionic hydrides must be stored in environments that rigorously exclude water and oxygen. The hydrides of rubidium and cesium, RbH and CsH, even ignite spontaneously in dry air.

The reaction of ionic hydrides with water can be represented by the following equations, where *M* represents a metal in either Group IA or IIA. The products are metal hydroxides and hydrogen.

Group IA ionic hydrides:

$$MH + H_2O \longrightarrow MOH + H_2$$

Group IIA ionic hydrides:

$$MH_2 + 2H_2O \longrightarrow M(OH)_2 + 2H_2$$

The rates of these reactions vary widely. Sodium hydride, for example, reacts with water even more violently than does sodium metal. Calcium hydride, however, reacts so moderately that its reaction with water is a small-scale source of hydrogen in the lab.

Covalent hydrides include some of our most vital compounds, such as water, ammonia, and methane.

Covalent Hydrides The elements that form the covalent hydrides are located on the right side of the periodic table. Even the *metallic* elements in Groups IIIA, IVA, VA, and VIA form covalent hydrides.

In covalent hydrides the bond that holds H varies widely in polarity according the electronegativity of the atom holding H. (See the table in the margin.) The electronegativities of the second elements in the covalent hydrides vary enough to put a $\delta+$ on H in some, a nearly zero partial charge in others, and a $\delta-$ charge in still others. The differences determine whether covalent hydrides react as donors of H^+, H, or H^-. Let's examine these differences more closely.

Electronegativities (H = 2.1)

IIIA	IVA	VA	VIA	VIIA
B	C	N	O	F
2.0	2.5	3.1	3.5	4.1
Al	Si	P	S	Cl
1.5	1.8	2.1	2.4	2.9
Ga	Ge	As	Se	Br
1.8	2.0	2.2	2.5	2.8
In	Sn	Sb	Te	I
1.5	1.7	1.8	2.0	2.2

The binary hydrides with $\delta+$ on H are *acids* and so are proton donors. They include the hydrides of the halogens, Group VIIA (HF, HCl, HBr, HI), and Group VIA (H_2O, H_2S, H_2Se, H_2Te), the latter set being weaker proton donors than the hydrides of the halogens.

The difference in electronegativity between C (or any Group IVA element) and H is very small.

When the partial charge on H in a binary hydride is zero or nearly so, as it is in CH_4 (methane), we have neither an acid nor a base. Instead, hydrocarbons like methane tend to react as donors of hydrogen *atoms,* doing so by free radical chain mechanisms of the type introduced in Special Topic 14.1.

Covalent hydrides with a $\delta-$ charge on H tend to be *donors* of hydride ion, H^-. These occur where H is covalently joined to an atom of *weaker* electronegativity than itself, such as those of some of the representative metallic elements as well as some metalloids. Thus, the hydrides of boron, aluminum, arsenic, and tin are donors of H^-. Although they are not ionic hydrides with preformed H^- ions, they still supply H^- in chemical reactions and thus are reducing agents. Some hydrides are so reactive toward oxygen that they burst into flame in air. Thus, diborane, B_2H_6, the simplest hydride of boron, is a gas that must be handled out of contact with air; it reacts spontaneously with oxygen as follows.

For explaining the unusual structures of the hydrides of boron, William Lipscomb received the 1976 Nobel Prize in chemistry.

$$\underset{\text{diborane}}{B_2H_6(g)} + 3O_2(g) \longrightarrow \underset{\text{boron oxide}}{B_2O_3(s)} + 3H_2O$$

Transition Metal Hydrides The hydrides of the transition metals have an unusually complex chemistry. Some have definite formulas like the hydrides of nickel (NiH_2), iron (FeH_2), and uranium (UH_3), but other transition metal hydrides have no definite formulas. They appear to be solutions of hydrogen in the metal, in which hydrogen is lodged within the interstices between atoms, and the material retains the electrical conductivity of a metal. Such hydrides often behave as donors of H_2, rather than donors of H^+ or H^-. We can summarize the binary hydrides as follows.

Types of Binary Hydrides

H^- donors	Ionic hydrides and covalent hydrides with $\delta-$ on H (generally metal hydrides)
$H\cdot$ donors	Compounds of carbon and hydrogen with C—H bonds having no partial charge on H, like CH_4
H^+ donors	Covalent hydrides with $\delta+$ on H (generally nonmetal hydrides)
H_2 donors	Some transition metal hydrides

Uses of Hydrogen as a Chemical Most manufactured hydrogen is used immediately, without isolation, to make ammonia. Purified hydrogen is also used to manufacture oleomargarine and other hydrogenated vegetable oils like Crisco and Spry.

Hydrogen is also a rocket fuel. It is loaded into fuel tanks as a liquid and then allowed to mix and react with oxygen in the propulsion chamber. There it burns to give hot, expanding steam whose pressure provides the force of propulsion.

Liquid hydrogen is used as a rocket fuel.

1.3 OXYGEN

The members of the **oxygen family,** given in Table 1.2, form hydrides and oxides with similar formulas. Our study focuses on the first two members, oxygen in this chapter and sulfur in the next.

Elemental Oxygen Oxygen is a colorless, odorless, tasteless gas that in its most stable state under ordinary conditions exists almost entirely as diatomic molecules, O_2. Oxygen has three isotopes, with oxygen-16 making up nearly 99.8% of the total. Both liquid and solid oxygen are light blue in color, and both are paramagnetic (see Figure 1.2).

Isotope	% Abundance
^{16}O	99.759
^{17}O	0.0374
^{18}O	0.2039

TABLE 1.2 The Oxygen Family—Group VIA

Element	Atomic Symbol	Melting Point (°C)	Boiling Point (°C)	Appearance[a]
Oxygen	O	−218	−183	Colorless gas
Sulfur	S	113[b]	445	Yellow, brittle solid
Selenium	Se	217	685	Bluish-gray metal
Tellurium	Te	452	1390	Silvery-white metal
Polonium	Po	254	962	Metallic (intensely radioactive)

Some Compounds

Hydrides		Oxides	
Formula	Boiling Point (°C)	Formula	Boiling Point (°C)
H_2O	100	O_3 (ozone)[c]	−112
H_2S	−61	SO_2	−10
		SO_3	−45
H_2Se	−42	SeO_2	Sublimes; MP > 300
		SeO_3	MP 118
H_2Te	−2	TeO_2	Decomposes
		TeO_3	Decomposes
—	—	PoO_2	Decomposes

[a] At room temperature and atmospheric pressure.

[b] For orthorhombic sulfur when heated rapidly.

[c] An allotrope of oxygen, not a compound, but it can also be thought of as an oxide of oxygen.

Sources of Oxygen

Oxygen makes up 20.95% (v/v) of dry air and has been at this concentration for an estimated 50 million years.

The great French scientist, Antoine Laurent Lavoisier (1743–1794), was the first to recognize oxygen as an *element,* although others had prepared it earlier—Carl Wilhelm Scheele, a Swedish chemist (1742–1786), and Joseph Priestly, an English scientist (1733–1804).

Industrial Production and Uses of Oxygen Oxygen is obtained in commercial quantities today by the careful distillation of liquid air. Of the chief components of air, nitrogen boils at 77.4 K, argon at 87.5 K, and oxygen at 90.2 K. Hence, nitrogen and argon distill from liquid air first, leaving the

(*a*)

(*b*)

FIGURE 1.2

Liquid oxygen. (*a*) As a pale blue liquid (BP, − 183 °C). (*b*) Because oxygen is paramagnetic, its molecules having two unpaired electrons, liquid oxygen is held between the opposite faces of a magnet.

oxygen. Oxygen usually ranks among the top five chemicals produced annually in the United States; in the mid 1990s, the annual U.S. production was over 23 million tons (6.6×10^{11} moles).

Over 50% of the manufactured oxygen is consumed by the production of metals, mostly in the basic oxygen process for making steel. Another 20% is used in the chemicals industry to make various oxygen derivatives of hydrocarbons, including plastics. The treatment of sewage, wastewater, and waste paper pulp fluids; medical uses; and rocketry consume most of the remaining oxygen production (see Figure 1.3).

Small-Scale Preparation of Oxygen Small amounts of oxygen can be made by heating potassium chlorate in the presence of manganese dioxide.

$$2KClO_3(s) \xrightarrow[150-200\ ^\circ C]{MnO_2} 3O_2(g) + 2KCl(s)$$

The manganese dioxide serves as a catalyst; without it a much higher temperature is needed. For virtually all laboratory needs, however, oxygen is obtained as a compressed gas from commercial sources.

Oxygen Formation and Consumption by Natural Processes The production and consumption of oxygen in nature, which are in almost perfect balance, occur as a complex cycle called the *oxygen cycle* (see Special Topic 1.1). Although several natural processes consume oxygen—respiration, decay, combustion, and other oxidations—it is replenished just as rapidly by **photosynthesis.** Plants carry on photosynthesis when the molecules of chlorophyll in green leaves trap electromagnetic energy from the sun and use it to convert carbon dioxide and water into complex compounds, like carbohydrates. Oxygen is the other chief product of photosynthesis. The overall,

Chlorophyll is responsible for the green color of leaves, needles, and some barks.

FIGURE 1.3

Liquid oxygen, LOX, was the oxidizing agent that reacted with the fuel to give the energy for propelling American astronauts to the moon and back.

SPECIAL TOPIC 1.1 / GEOCHEMICAL CYCLES IN NATURE

A **geochemical cycle** consists of all of the reactions in nature known to contribute to the consumption and regeneration of a specific substance. Both oxygen and nitrogen are major constituents of air, and each element is used up and remade on such a continuous and equal basis that the concentrations of these elements in air has not changed since measurements began.

OXYGEN CYCLE

Some of the chief features of the oxygen cycle in nature are shown in Figure 1 together with estimates of the quantities of materials that participate, given in units of 10^{12} mole per year. (For example, the estimated amount of oxygen tied up in "organisms, land" is $38{,}000 \times 10^{12}$ mol.) Biological *reduction* in the figure refers chiefly to photosynthesis, which occurs among both land and sea plants, particularly phytoplankton in the oceans, the latter accounting for over half of all the Earth's photosynthetic acitivty. Despite the seemingly huge quantities of materials that participate in the oxygen cycle, the actual solar energy consumed by photosynthesis amounts to only about 0.04% of the solar energy received by the Earth.

Biological oxidation in Figure 1 refers to the respiration of plants and animals, and to combustion and decay. It includes the conversion of N and S atoms among biological molecules to nitrates and sulfates, as well as the slow *chemical oxidation* of iron. Also parts of the oxygen cycle are the formation of carbonate rocks (*lithification* in the figure) and the thermal decomposition of the rocks in volcanic magma (*magmafication*). As *vulcanism* occurs, oxygen is discharged to the atmosphere along with other gases, like CO_2 and SO_3 (from the thermal decomposition of various types of rocks).

Fortunately, the oxygen cycle gives us air with only 21% O_2. One scientist has estimated that if the air were only modestly richer in oxygen, 30% instead of 21%, no forest fire could be put out. Rates of reactions, you will recall, can be accelerated simply by using higher concentrations of reactants.

NITROGEN CYCLE

By complex reactions called *nitrogen fixation,* microorganisms take nitrogen from the air and make compounds they need, in-

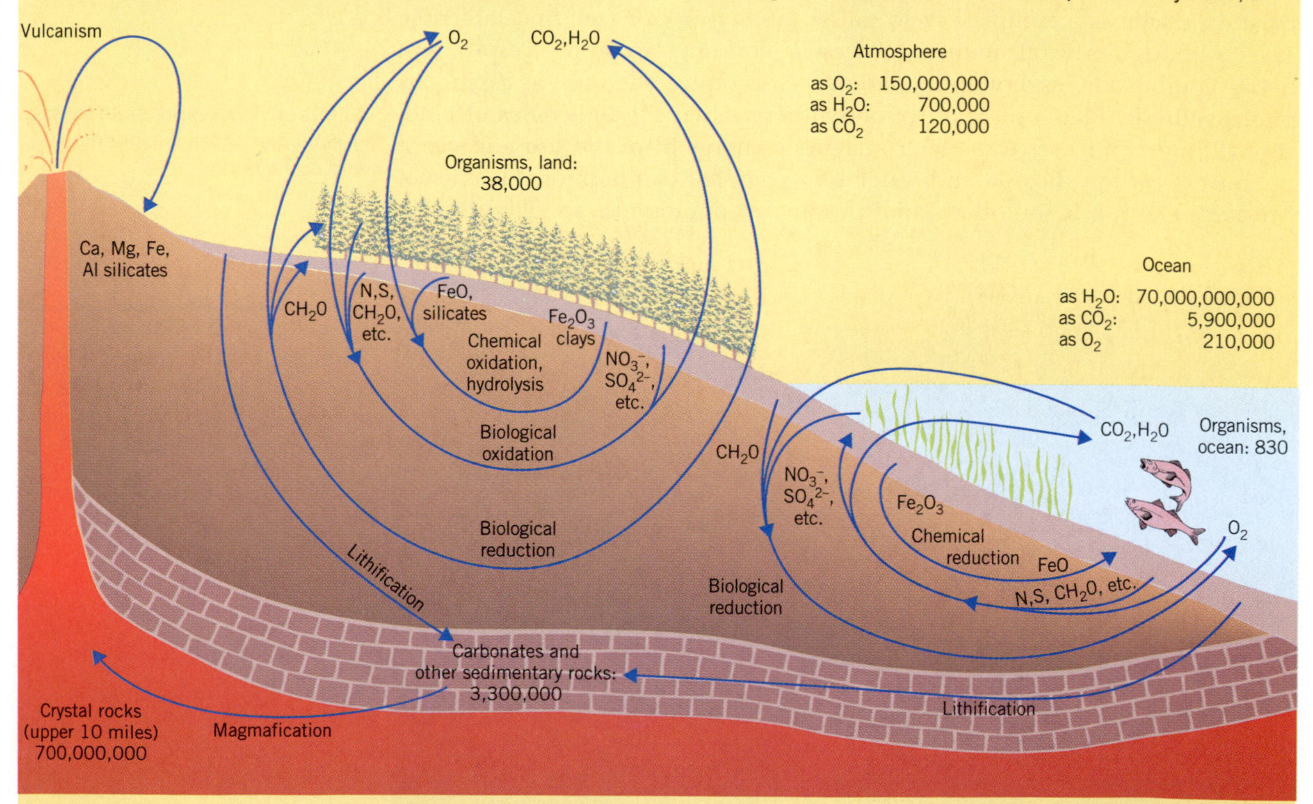

FIGURE 1

The oxygen cycle in nature involves air, land, and sea. (From A. N. Strahler and A. H. Strahler, *Introduction to Environmental Science,* copyright © 1974, Hamilton Publishing Co., Santa Barbara. Used by permission.)

cluding ammonia. Eventually, the nitrogen compounds in living things are reconverted to nitrogen by decay, so a huge *nitrogen cycle* exists in nature (Figure 2). In it, nitrogen nuclei move from the air through the biosphere and back again. (The *biosphere* is the sum of all parts of the Earth that have living systems.) The numbers on the curved arrows in Figure 2 are the estimated flows of nitrogen in units of 10^{12} moles per year. (For example, *biological fixation* is estimated to convert 3.9×10^{12} mol per year of atmospheric nitrogen into nitrogen compounds.) The numbers that stand apart from curved arrows are estimates of the amounts of nitrogen present in one form or another in each region in units of 10^{12} moles. (For example, the atmosphere is estimated to hold $270{,}000{,}000 \times 10^{12}$ mol of nitrogen.) *Atmospheric fixation* in Figure 2 means the chemical reactions, some initiated by lightning discharges, that make nitrogen oxides. *Industrial fixation* refers largely to the Haber–Bosch process. *Denitrification* is the oxidation of nitrogen in plant and animal materials, most of it in the −3 oxidation state, into nitrogen and its oxides.

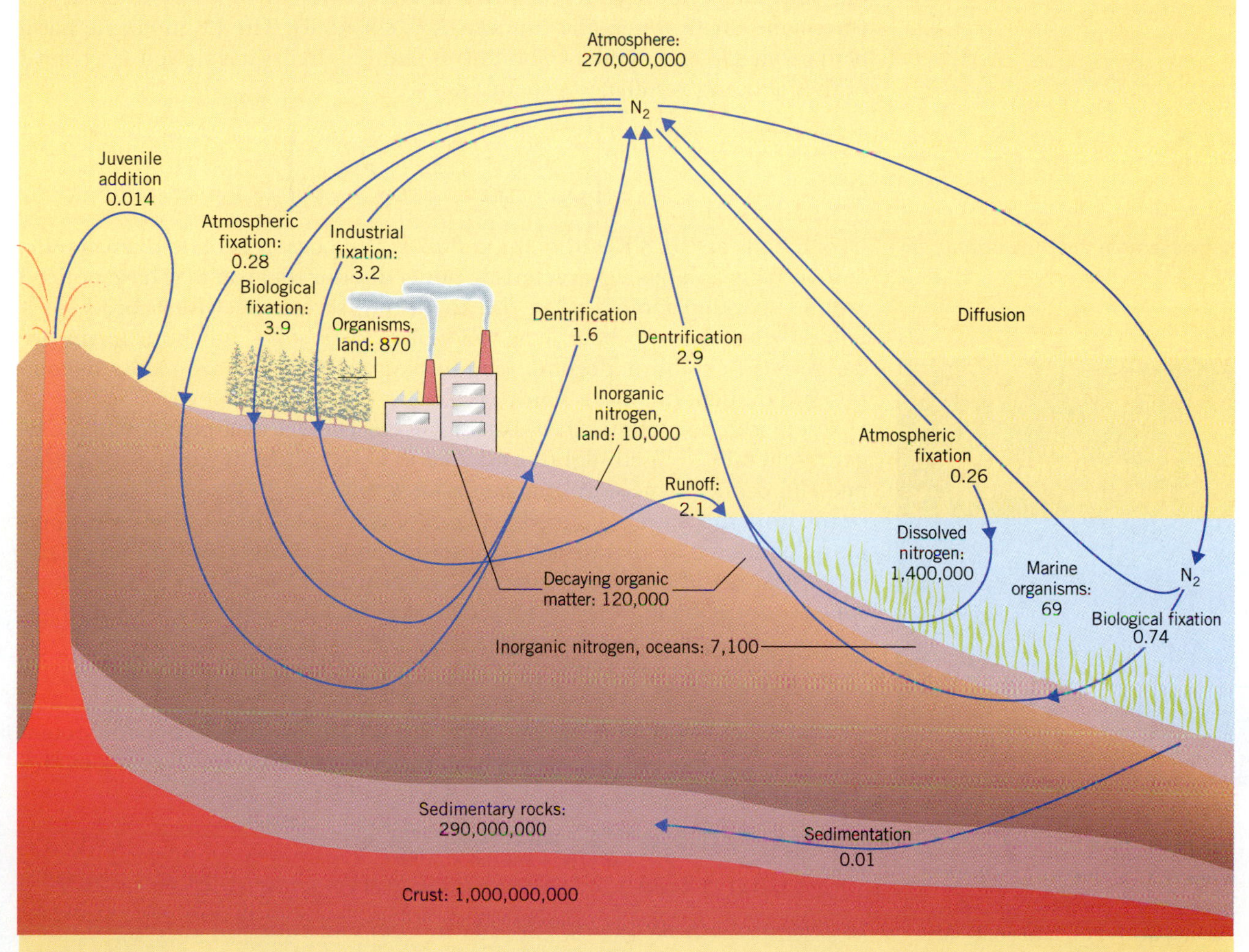

FIGURE 2

The nitrogen cycle in nature. (From A. N. Strahler and A. H. Strahler, *Introduction to Environmental Science,* copyright © 1974, Hamilton Publishing Co., Santa Barbara. Used by permission.)

multistep process can be summarized as follows where the formula (CH_2O) represents the *empirical* formula of one of the major products, glucose, $C_6H_{12}O_6$, or $(CH_2O)_6$.

$$n\mathrm{CO_2} + n\mathrm{H_2O} \xrightarrow[\text{Several steps}]{\text{Photosynthesis}} (\mathrm{CH_2O})_n + n\mathrm{O_2}$$

The greatest users of oxygen, as we said, are natural processes—respiration, combustion, and decay—any of which could be represented by the reverse of the equation for photosynthesis. They are all oxidative processes, converting the carbon and hydrogen combined in organic substances back to carbon dioxide and water.

Ozone

Ozone is triatomic oxygen, O_3. Different molecular (or crystalline) forms of the same element are called **allotropes,** so ozone is an allotrope of dioxygen (the name for O_2 when absolute clarity is essential). The O_3 molecule has a bent geometry, and in valence-bond (resonance) theory it is viewed as a hybrid of the following resonance structures.

$$\left[\ddot{\mathrm{O}}{-}\ddot{\mathrm{O}}{=}\ddot{\mathrm{O}} \longleftrightarrow \ddot{\mathrm{O}}{=}\ddot{\mathrm{O}}{-}\ddot{\mathrm{O}} \right]$$

Ozone is from the Greek *ozein,* "to smell."

Ozone boils at 161.3 K and melts at 80.7 K. As a gas, it is blue, diamagnetic, and unstable and has a characteristic pungent odor that you may have noticed around sparking motors or in the air during severe electrical storms. As a liquid, ozone is deep blue, it is violet-black as a solid, but both forms are dangerously explosive, decomposing to oxygen. (Gaseous ozone also decomposes to oxygen, but more slowly.)

Ozone can be synthesized by passing an electrical discharge through dioxygen. Generally, oxygen slightly enriched in O_3 is sufficient for any laboratory need for ozone. Some sterilizing lamps work by making a low concentration of ozone from the action of ultraviolet light on the O_2 in air. Worldwide, ozone's most common commercial use is to sterilize water. Many cities in Europe use ozone instead of chlorine for water treatment. Ozone occurs in the stratospheric ozone shield, where it is beneficial to us, forming by a chemical chain reaction as discussed in *Chemicals in Use* 7 (see footnote 1).

FIGURE 1.4
The damage to these trees near Lake Tahoe, California, was caused by a combination of ozone, acid rain, and road salt spray.

Ozone as an Oxidizing Agent Ozone has high standard reduction potentials in both acidic and basic media, which indicates that O_3 is a strong oxidizing agent.

$$\mathrm{O_3}(g) + 2\mathrm{H^+}(aq) + 2e^- \rightleftharpoons \mathrm{O_2}(g) + \mathrm{H_2O} \qquad E^\circ = +2.08\ \mathrm{V}$$

$$\mathrm{O_3}(g) + \mathrm{H_2O} + 2e^- \rightleftharpoons \mathrm{O_2}(g) + 2\mathrm{OH^-}(aq) \qquad E^\circ = +1.24\ \mathrm{V}$$

In acid, ozone is second only to fluorine (among common chemicals) as an oxidizing agent, and it is sometimes used as an oxidant to make compounds. Ozone attacks just about anything, including trees (see Figure 1.4), fabrics, dyes, rubber, and lung tissue, making ozone one of the most serious pollutants in smog (see *Chemicals in Use* 8). Exposure to ozone at a level of 1.0 $\mu g\ m^{-3}$ (1.0 ppm) for over 10 minutes is dangerous for all people. At a level of only

In severe smog episodes, the ozone level can reach 1 ppm.

0.5 μg m^{-3} (0.5 ppm) of ozone, the physical activities of children and the elderly must be curtailed to reduce their inhaling ozone deeply.

Basic and Ionic Oxides

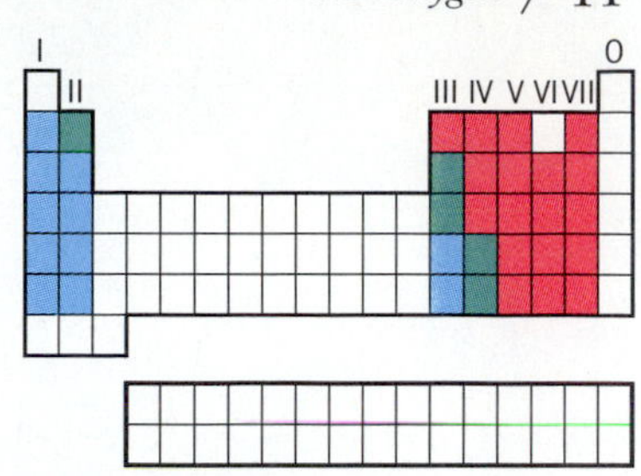

Among the representative elements:
- Form basic oxides
- Form amphoteric oxides
- Form acidic oxides

Except for the first four noble gases—helium, neon, argon, and krypton—all elements form oxides, many being preparable directly from the elements. The figure in the margin indicates which of the representative elements have basic oxides, which are acidic, and which are amphoteric.

We learned about *basic oxides* or *basic anhydrides* in Section 4.3. Most basic oxides are ionic compounds, particularly those of Group IA metals as well as of calcium, strontium, and barium in Group IIA. What makes these metal oxides basic is the oxide ion, O^{2-}, a powerful proton acceptor that reacts vigorously with water to give the hydroxide ion.

$$O^{2-} + H_2O \longrightarrow 2OH^-$$

The few basic oxides that are quite soluble in water, like Na_2O, dissolve by this reaction, and the actual solute becomes the corresponding water-soluble hydroxide. Sodium oxide, for example, dissolves as it reacts with water to form a solution of sodium hydroxide.

$$Na_2O(s) + H_2O \longrightarrow 2NaOH(aq)$$

Like all compounds that react with water, Na_2O can remove water from humid air, so it must be stored in tightly capped bottles.

All of the other Group IA metal oxides react with and dissolve in water by a similar reaction so all must be stored in a dry atmosphere. The oxide ion cannot exist in solution in water. *The strongest base that can be in an aqueous solution is the hydroxide ion.* All stronger bases, like O^{2-}, react with water and are replaced by the OH^- ion.

The Group IIA metal oxides, except for beryllium oxide, are also basic. They all melt above 1900 °C, which indicates that these are also *ionic* compounds containing the oxide ion. Magnesium oxide, MgO, is typical of the Group IIA metal oxides, except for that of beryllium. (Beryllium oxide, while having a high melting point, has bonds that are mostly covalent in nature.) Although MgO is insoluble in water, it does react with water to form the relatively *insoluble* hydroxide. In acids, however, MgO dissolves by the following reaction.

Group IIA Oxide	Melting Point (°C)
BeO	2530
MgO	2800
CaO	2600
SrO	2430
BaO	1923

$$MgO(s) + 2H^+(aq) \longrightarrow Mg^{2+}(aq) + H_2O$$

The oxides of the Group IIA metals are *refractory* materials, meaning that they can withstand high heat in the presence of air and yet still conduct heat well. They melt so high that some are used to make the bricks that line ovens and furnaces where pottery and china are baked.

Covalent Metal Oxides

Once we leave the oxides of the Group IA and IIA metals and look at the oxides of other metals, we find a mixed situation. Some transition metal oxides have low melting points, suggesting that they consist of relatively nonpolar molecules. Examples are dimanganese heptoxide (Mn_2O_7, MP 6 °C), ruthenium tetroxide (RuO_4, MP 25 °C), and osmium tetroxide (OsO_4, MP 40 °C). Let us consider why these metal oxides are *covalent* in character, rather than ionic.

Interestingly, osmium metal has the fourth highest melting point of all the elements—3040 °C—but it has an oxide that is quite low melting.

The oxidation numbers of the metals in the low-melting covalent metal oxides are particularly high: +7 for Mn in Mn_2O_7, +8 for both Ru in RuO_4 and Os in OsO_4. If Os actually had to exist as a *cation* in OsO_4, its charge would be 8+. In other words, the osmium center in OsO_4 would have a very high positive charge density. With a positive charge density this high, the hypothetical Os^{8+} core in OsO_4 would strongly polarize electron clouds and force electrons from any chemical species next to it and so fashion a covalent bond to the neighbor. Moreover, Os^{8+} would have a much smaller radius than any osmium cation of lower charge because the radii of cations shrink as their positive charges increase. Small size and high charge prevent Os^{8+} ions from existing together with oxide ions; instead, a *sharing* of electrons is forced, and *covalent* bonds from oxygen to osmium are present. The bonding situation in OsO_4 illustrates a general rule.

The higher the oxidation state of the metal atom in its compounds, the more covalent and less ionic are its bonds.

As the oxidation number of the cation *decreases,* the bonds to it from adjacent atoms become less covalent and more ionic. We see this as we move from the +8 state of osmium to its +4 state in osmium dioxide, OsO_2. The two crystalline forms of osmium dioxide, OsO_2, do *not* have low melting points, like the tetroxide, suggesting that the bonds in OsO_2 are ionic. Between 350 °C and 400 °C, one oxide changes to the other and at 500 °C, the latter decomposes and does not melt. As we said, this behavior suggests ionic bonds, not covalent bonds, which indicates that Os^{4+} is much less able than Os^{8+} to distort electron clouds and force covalent bonds to form.

TiO_2 is the most commonly used white pigment in paints.

The oxidation states of the metals are also +4 in thorium dioxide, ThO_2 (MP 2950 °C), and in titanium dioxide, TiO_2 (MP 1840 °C). The high melting points suggest that the bonds in these oxides are more ionic than covalent. (We can only say "suggest," not "prove." We will see why soon.)

Effect of a Central Metal Ion's Oxidation State on an Oxide's Acidity or Basicity

The oxidation state of the metal in a metal oxide also influences the acid–base properties of the compound.

Acidic Metal Oxides Oxides of metals in high oxidation states, like Cr in chromium(VI) oxide, CrO_3, tend to be acidic oxides. CrO_3 melts at 197 °C, which is low enough to suggest that its bonds are covalent in character. CrO_3 is the acidic anhydride of chromic acid, H_2CrO_4, which is able to exist in this molecular form in an aqueous solution at very low pH (<1). These properties make up part of the evidence for the following important generalization.

As a general rule, the higher the oxidation state of the metal in a metal oxide, the more acidic is the oxide.

Osmium tetroxide, with the high +8 oxidation state for Os, is another acidic oxide illustrating this generalization. In concentrated potassium hydroxide, for example, OsO_4 neutralizes hydroxide ion according to the following net ionic equation.

$$OsO_4(s) + 2OH^-(aq) \longrightarrow \underset{\text{perosmate ion}}{[OsO_4(OH)_2]^{2-}(aq)}$$

Amphoteric Metal Oxides Oxides of metals with lower oxidation numbers are not acidic but are either amphoteric or basic. An *amphoteric oxide* is one that can neutralize either acid or base. Aluminum oxide is a common example. Because this oxide is very high melting (≈ 2000 °C), its bonds are believed to be essentially ionic, not covalent, as they are in the much lower melting CrO_3. Thus, we can regard Al_2O_3 as consisting of Al^{3+} ions and O^{2-} ions.

Amphoteros is Greek for "partly one and partly the other."

Because of the powerful basicity of its oxide ion, aluminum oxide neutralizes acids by the following reaction.

$$Al_2O_3(s) + 6H^+(aq) + 3H_2O \longrightarrow 2[Al(H_2O)_6]^{3+}(aq)$$

Al_2O_3 also neutralizes a strong base, like OH^-.

$$Al_2O_3(s) + 2OH^-(aq) + 7H_2O \longrightarrow 2[Al(H_2O)_2(OH)_4]^-(aq)$$

The actual product of the reaction of Al_2O_3 with OH^- varies somewhat with how basic the solution is made.

This reaction is explained by the acidity (in a Lewis acid sense) of the Al^{3+} cation in Al_2O_3. Lewis acids are strong electron-pair acceptors (see Section 7.9), and Al^{3+} in Al_2O_3 has a high positive charge density, enabling it to attract species with electron pairs, like OH^- or H_2O.

Chromium also has an amphoteric oxide, Cr_2O_3, but yet another oxide, CrO, is basic. In Cr_2O_3, Cr is in a +3 oxidation state, and in CrO it is in the +2 state. Thus, the three oxides of chromium that we have mentioned, CrO_3, Cr_2O_3, and CrO, respectively acidic, amphoteric, and basic oxides, illustrate a general trend.

The oxides of transition metals become more basic (less acidic) as the oxidation number of the metal ion decreases.

Basic Metal Oxides CrO is more basic (less acidic) than Cr_2O_3 because of the lower oxidation state of Cr in CrO. The change of chromium from the +3 state in Cr_2O_3 to the +2 state in CrO makes the metal ion in CrO a *weaker* Lewis acid. Yet this does not weaken the basicity of the O atoms. Quite the contrary; they become *stronger* bases. The decrease in oxidation number of the metal ion allows the electron pairs on the O atoms to experience *less* attraction toward the cation. Therefore, electron pairs on an oxygen atom bound to a metal ion of low oxidation number are *more* available to help the O atoms function as electron-pair donors, or Lewis bases. Illustrating this generalization are the powerfully basic oxide ions found among the Group IA and Group IIA metal oxides, where the metal ions have low oxidation numbers. Similarly, many transition metals in the +2 oxidation state also form basic oxides, like NiO and FeO.

A few metal oxides, for example, MnO_2 and PbO_2, are inert both to acids and bases (unless the reaction is a redox reaction, not an acid–base reaction).

Acidity of Nonmetal Oxides Most of the oxides of the nonmetals are acidic oxides (acidic anhydrides), as was mentioned in Section 4.3. Examples are those of nitrogen, sulfur, and the halogens, as well as carbon dioxide.

Nitrogen has similar acidic anhydrides, which will be discussed in the next section.

Among the oxides of sulfur, SO_3 is much more acidic than SO_2, which parallels the difference in the oxidation numbers of S, namely, + 6 in SO_3 and + 4 in SO_2. SO_3, therefore, is analogous to CrO_3, which is also an acidic oxide whose central atom is in the + 6 state. SO_3 reacts with water to give the *strong* acid, sulfuric acid (H_2SO_4). In contrast, SO_2 interacts with water to give a much weaker acidic system, sulfurous acid, H_2SO_3 (or $SO_2 \cdot nH_2O$).

Hydrogen Peroxide and Its Salts

Hydrogen peroxide, H_2O_2, is a colorless liquid that freezes at −0.43 °C, boils at 150.2 °C, and has a density of 1.4425 g mL^{-1} (25 °C). It must be handled extremely carefully, being particularly dangerous when pure.

The H_2O_2 molecule is not planar but is in a twisted conformation, called a *skew-chain* structure, as the figure in the margin shows. This shape results from the repulsions of the electron clouds of the two unshared electron pairs on each O atom as well as those in the O—H bonds.

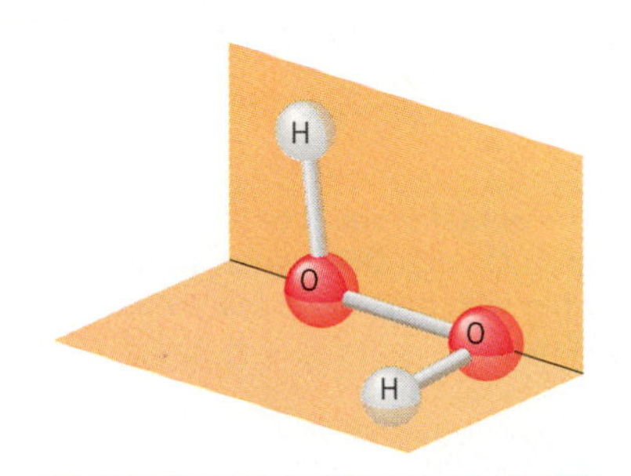

Hydrogen peroxide (solid phase)

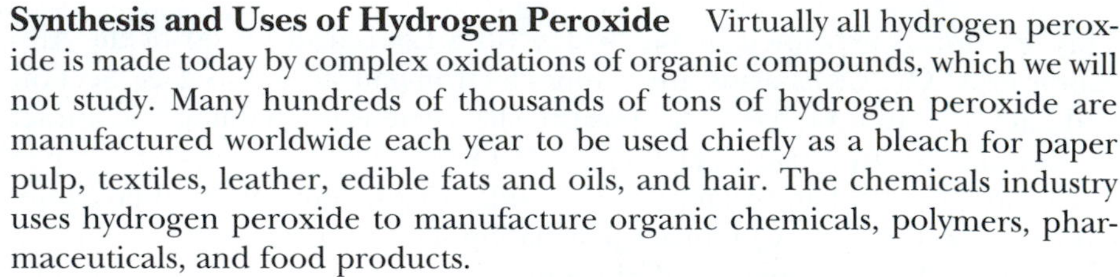

Synthesis and Uses of Hydrogen Peroxide Virtually all hydrogen peroxide is made today by complex oxidations of organic compounds, which we will not study. Many hundreds of thousands of tons of hydrogen peroxide are manufactured worldwide each year to be used chiefly as a bleach for paper pulp, textiles, leather, edible fats and oils, and hair. The chemicals industry uses hydrogen peroxide to manufacture organic chemicals, polymers, pharmaceuticals, and food products.

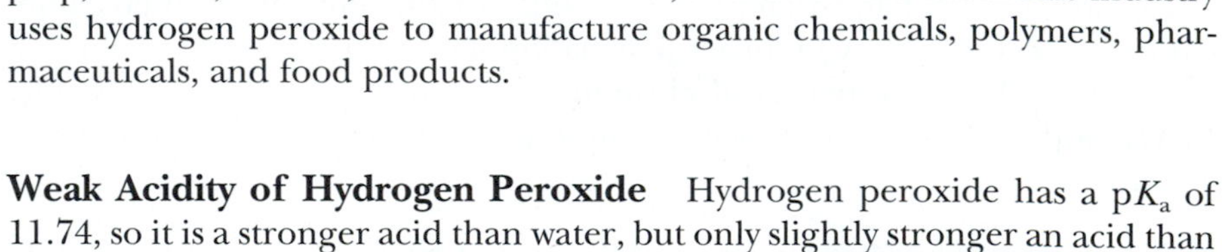

Weak Acidity of Hydrogen Peroxide Hydrogen peroxide has a pK_a of 11.74, so it is a stronger acid than water, but only slightly stronger an acid than HPO_4^{2-}, for which the pK_a is 12.35. A high pH, therefore, is required if the following equilibrium is to favor the hydroperoxide ion, HO_2^-.

$$OH^-(aq) + H_2O_2(aq) \rightleftharpoons H_2O + HO_2^-(aq)$$

As a proton acceptor, hydrogen peroxide is about a million times weaker than water, so the following equilibrium strongly favors the reactants.

$$H_3O^+(aq) + H_2O_2(aq) \rightleftharpoons H_2O + H_3O_2^+(aq)$$

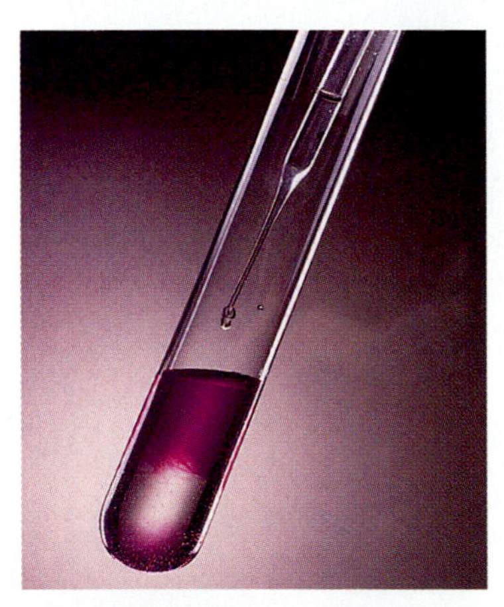

FIGURE 1.5

The permanganate ion is a strong enough oxidizing agent to oxidize hydrogen peroxide. In the region where the hydrogen peroxide has settled (prior to mixing), the purple color of the permanganate ion is destroyed as Mn^{2+} forms and small bubbles of oxygen evolve.

Redox Properties of Hydrogen Peroxide The following equations summarize the chief redox properties of the hydrogen peroxide/hydroperoxide ion system in acid and base.

$$H_2O_2(aq) + 2H^+(aq) + 2e^- \rightleftharpoons 2H_2O \qquad E° = +1.77\text{ V} \quad (1.1)$$

$$O_2(g) + 2H^+(aq) + 2e^- \rightleftharpoons H_2O_2(aq) \qquad E° = +0.69\text{ V} \quad (1.2)$$

$$HO_2^-(aq) + H_2O + 2e^- \rightleftharpoons 3OH^-(aq) \qquad E° = +0.87\text{ V} \quad (1.3)$$

Thus, in either an acidic solution (Equation 1.1) or a basic solution (Equation 1.3), hydrogen peroxide is a strong oxidizing agent. The *rates* of oxidation are higher in basic media, however.

Hydrogen peroxide will operate as a reducing agent—using the *reverse* of Equation 1.2 as the half-cell reaction—only toward very strong oxidizing agents, like MnO_4^- (see Figure 1.5). For example, H_2O_2 is able to reduce the

Mn in the permanganate ion from its + 7 state to its + 2 state by the following equation.

$$2MnO_4^-(aq) + 5H_2O_2(aq) + 6H^+(aq) \longrightarrow 2Mn^{2+}(aq) + 8H_2O + 5O_2(g)$$

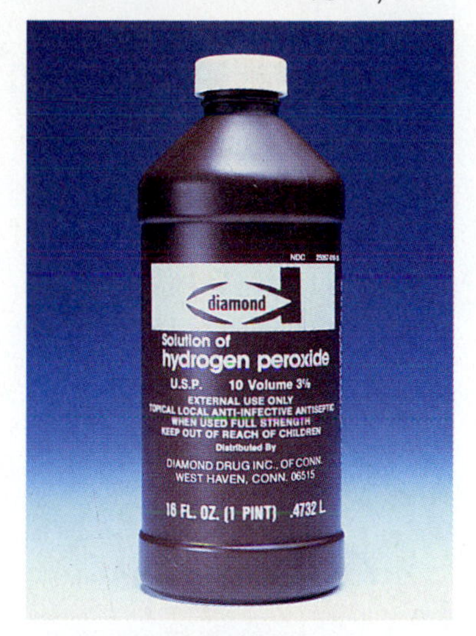

Dilute hydrogen peroxide

Safe Handling—The Decomposition of Hydrogen Peroxide After standing a long time, hydrogen peroxide eventually decomposes to water and oxygen.

$$2H_2O_2(l) \longrightarrow 2H_2O + O_2(g)$$

The decomposition is accelerated by almost anything, particularly if traces of base are present—by heavy metal ions, dirt, heat, and even water and oxygen, which are the *products* of its decomposition. This is why pure hydrogen peroxide is so dangerous. Because oxygen is a product, the decomposition of hydrogen peroxide can start and sustain fires.

Even 30% aqueous hydrogen peroxide should be handled only by experienced personnel. Material of this concentration and higher is generally stored in bottles made of special plastic or in bottles coated on the inside with wax. This is because glass surfaces often have alkaline properties that accelerate the decomposition of H_2O_2. The "peroxide" sold in drugstores is only 2 to 3% in concentration.

If you have ever used dilute hydrogen peroxide to disinfect a wound, you no doubt noticed the foaming. Blood catalyzes the decomposition of H_2O_2, and the oxygen kills bacteria.

Salts of Hydrogen Peroxide Hydrogen peroxide is the "parent" compound of the **peroxides,** a family of substances with O—O bonds. Metal peroxides like Na_2O_2 are not, however, made directly from hydrogen peroxide. Their synthesis is discussed in Section 3.3 of this supplement.

As you might expect, the peroxide ion, O_2^{2-}, is a powerful base, so when sodium peroxide is added to water it reacts almost quantitatively as follows to give an alkaline solution of hydrogen peroxide (which then decomposes to water and oxygen as described earlier).

Per, as in "peroxide," indicates a higher proportion of oxygen than in the normal oxide.

$$Na_2O_2(s) + 2H_2O \longrightarrow 2NaOH(aq) + H_2O_2(aq)$$

Thus, solid metal peroxides capable of this kind of reaction provide safer convenient sources of bleaching and disinfecting power than aqueous hydrogen peroxide. The decomposition of sodium peroxide by a trace of water in the presence of an organic material, like sugar, can cause a dangerous fire in a compound that is otherwise difficult to ignite (see Figure 1.6).

Superoxides The *simple* oxides of the Group IA metals have the general formula, M_2O. Of these metals, only lithium reacts with the oxygen in air to give mostly its simple oxide, Li_2O. (Some Li_2O_2 also forms.)

The simple oxides of Na, K, and Rb, namely, Na_2O, K_2O, and Rb_2O, are made by indirect methods.

$$4Li(s) + O_2(g) \longrightarrow \underset{\text{lithium oxide}}{2Li_2O(s)}$$

Sodium, on the other hand, reacts with the oxygen in air to give mostly its peroxide, Na_2O_2. (Some Na_2O also forms.)

$$2Na(s) + O_2(g) \longrightarrow \underset{\text{sodium peroxide}}{Na_2O_2}$$

The remaining alkali metals, potassium, rubidium, and cesium, react with oxygen in air to give **superoxides** with the general formula MO_2. We can illustrate the reaction with potassium.

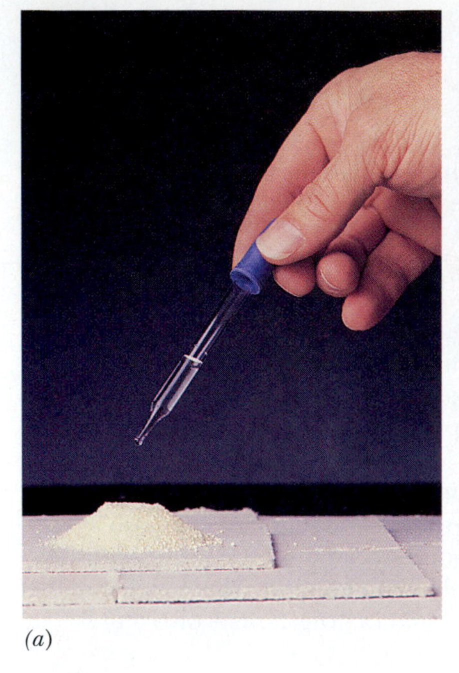

(*a*) (*b*)

FIGURE 1.6

Sodium peroxide as a strong oxidizing agent. (*a*) A drop of water is about to be added to a mixture of sodium peroxide (Na_2O_2) and ordinary sugar. (*b*) The hydrolysis of the peroxide is so exothermic that it ignites the mixture. Sodium peroxide is a powerful oxidizing agent and causes the sugar to burn vigorously.

$$K(s) + O_2(g) \longrightarrow \underset{\text{potassium superoxide}}{KO_2(s)}$$

The superoxides are crystalline, saltlike compounds of the superoxide ion, O_2^-, which is stable only in the presence of the larger Group IA and IIA cations, like K^+, Rb^+, Cs^+, Sr^{2+}, and Ba^{2+}. The O_2^- anion has one unpaired electron, so the superoxides are paramagnetic. The anion is unstable toward water, and the reaction of O_2^- with water liberates oxygen.

$$2O_2^-(s) + H_2O \longrightarrow O_2(g) + HO_2^-(aq) + OH^-(aq) \qquad (1.4)$$

The hydroperoxide ion, HO_2^-, that also forms decomposes to give more oxygen.

$$2HO_2^-(aq) \longrightarrow 2OH^-(aq) + O_2(g)$$

The superoxides, therefore, must be kept away from water. On a large enough scale, the reaction with water can generate enough oxygen (and heat) to set fire to nearby combustible materials.

The superoxides also react with carbon dioxide to give carbonates and oxygen. With potassium superoxide, for example, we have

$$4KO_2(s) + 2CO_2(g) \longrightarrow 2K_2CO_3(s) + 3O_2(g)$$

This reaction can be used to remove CO_2 from air and replace it with O_2. Thus, in a closed chamber, like a spacecraft, this reaction can be used to keep the air fresh. Of course, such air also has moisture, so this water would also decompose KO_2, as shown by Equation 1.4. But one product in Equation 1.4 is the OH^- ion, which also combines with CO_2.

$$KOH(s) + CO_2(g) \longrightarrow KHCO_3(s)$$

$$OH^- + CO_2 \longrightarrow HCO_3^-$$

So CO_2 could be removed from either dry or humid spacecraft air by KO_2 and be replaced by oxygen.

1.4 NITROGEN

Isotope	% Abundance
^{14}N	99.63
^{15}N	0.37

Some properties of the elements of the **nitrogen family**—nitrogen, phosphorus, arsenic, antimony, and bismuth—are given in Table 1.3. We will study only two members in any detail—nitrogen here and phosphorus in the next chapter. Another of the important chemical cycles in nature, the *nitrogen cycle*, is described in Special Topic 1.1.

Sources and Chief Uses of Nitrogen Nitrogen is a colorless, odorless, and tasteless gas that makes up 78% (v/v) of dry air. Roughly 25 to 30 million tons (800 to 1000 billion moles) of nearly oxygen-free nitrogen is produced each year from liquified air. The largest single use of nitrogen (30%) are in an oil field operation called *enhanced oil recovery* where pressurized nitrogen gas forces oil from subterranean deposits. Another use of nitrogen is by the metals and chemicals processing industries to replace air when a blanketing atmosphere of an unreactive gas is needed. Companies that make electronics components and semiconductor chips use ultrapure nitrogen (99.9999% pure) for this purpose.

Nitrogen was discovered to be an element in 1772 by an English scientist, Daniel Rutherford (1749–1819).

Its low boiling point (77 K) and chemical unreactivity make liquid nitrogen an ideal and relatively inexpensive coolant in research. Manufacturers of frozen seafood, poultry, and meat products also use liquid nitrogen for fast freezing.

Oxidation States of Nitrogen One of the remarkable properties of nitrogen is its wide range of oxidation states, ten in all (see Table 1.4). They occur in all possible whole-number states from -3 to $+5$ as well as in the $-1/3$ state. Moreover, many nitrogen compounds have nitrogen in more than one state, as in NH_4NO_3, where the -3 state is in NH_4^+ and the $+5$ state is in NO_3^-.

Ammonia

The oxidation state of N in NH_3 is -3, the most reduced state. The pungent odor of ammonia can be detected at a level in air of only about 50 to 60 ppm. At levels of 100 to 200 ppm, ammonia sharply irritates the eyes and air passages to the lungs. At high concentrations, ammonia vapor makes the lungs fill with fluid, which can quickly cause death unless prompt aid is given.

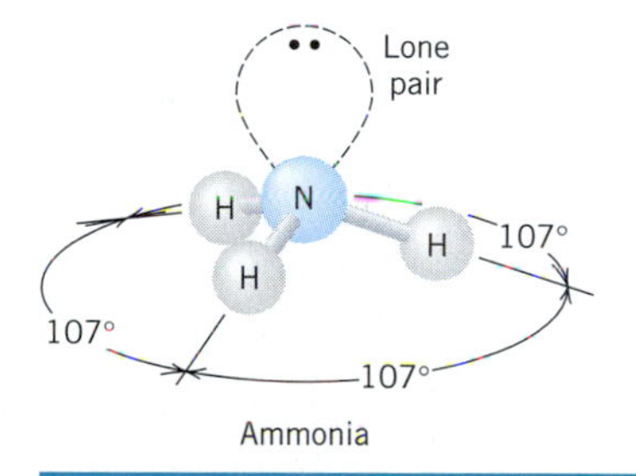

Ammonia structure

TABLE 1.3 The Nitrogen Family—Group VA

Name	Symbol	Melting Point (°C)	Boiling Point (°C)	Important Types of Compounds
Nitrogen	N	−210	−196	Nitrates (fertilizers, explosives) Oxides (air pollutants) Ammonia (fertilizer)
Phosphorus	P	44	281	Phosphates and polyphosphates (detergents, fertilizers)
Arsenic	As	815[a]	613[b]	Arsenates (pesticides)
Antimony	Sb	631	1750	Lead–antimony mixtures (alloys for batteries and type metal)
Bismuth	Bi	271	1560	Low-melting alloys for fire alarms and sprinkler systems

[a] Under 28 atm pressure.

[b] Sublimation temperature.

TABLE 1.4 Oxidation States of Nitrogen

Oxidation Number	Examples and Types: Formula	Name
−3	NH_3	Ammonia
	NH_4^+	Ammonium ion
	Mg_3N_2, Li_3N	Saltlike nitrides
	BN	Diamondlike nitrides
	$(CN)_2$	Covalent nitrides
−2	N_2H_4	Hydrazine
−1	NH_2OH	Hydroxylamine
−1/3	N_3^-	Azide ion
0	N_2	Nitrogen
+1	N_2O	Dinitrogen monoxide (nitrous oxide)
+2	NO	Nitrogen monoxide (nitric oxide)
+3	N_2O_3	Dinitrogen trioxide (nitrous anhydride)
+4	NO_2	Nitrogen dioxide
	N_2O_4	Dinitrogen tetroxide
+5	HNO_3	Nitric acid
	NO_3^-	Nitrate ion
	N_2O_5	Dinitrogen pentoxide (nitric anhydride)

Never inhale the vapors coming from a bottle of concentrated aqueous ammonia. Pour this liquid at the hood.

Haber–Bosch Synthesis of Ammonia Ammonia is made today by the **Haber–Bosch process** for which the central reaction of a multistep process is the combination of nitrogen and hydrogen (see *Chemicals in Use* 12).

$$N_2(g) + 3H_2(g) \rightleftharpoons 2NH_3(g)$$

In terms of *moles* manufactured, no compound surpasses ammonia.

Ammonia ranks in the top five of all chemicals made in the United States in terms of total mass—17 to 18 million tons (900 to 1000 billion moles) in the mid 1990s.

Ammonia, in one form or another, is the world's most important nitrogen fertilizer, about a quarter of all manufactured ammonia going directly for fertilizer (see Figure 1.7). The rest of the manufactured ammonia is used mostly to make chemicals, such as nitric acid, dyes, pharmaceuticals, and cleaning agents. Some ammonia is used as the heat-exchanger fluid in large refrigeration units.

FIGURE 1.7

Liquid ammonia is injected into the soil before planting to provide a nitrogen fertilizer. Although the liquid vaporizes in the soil, most of its molecules are trapped by dissolving in soil moisture.

Liquid ammonia must be handled at an efficient fume hood.

Physical Properties of Ammonia Ammonia (BP −33.4 °C; MP −77.7 °C) is easily liquified and has a relatively high heat of vaporization (1370 J g^{-1}). Liquid ammonia is purchased in steel cylinders that keep it under a pressure, and wide-mouthed vacuum bottles called *Dewar flasks* are generally employed as its containers in experimental work.

Ammonia's high solubility in water, 51.8 g NH_3/100 g H_2O at 20 °C, arises largely from the ability of its molecules to form hydrogen bonds to water molecules. Some dissolved NH_3 also reacts with water to give the following equilibrium.

$$NH_3(aq) + H_2O \rightleftharpoons NH_4^+(aq) + OH^-(aq)$$

Ammonia's solubility in water creates a partial vacuum in an interesting lecture demonstration, the ammonia "fountain" (see Figure 1.8).

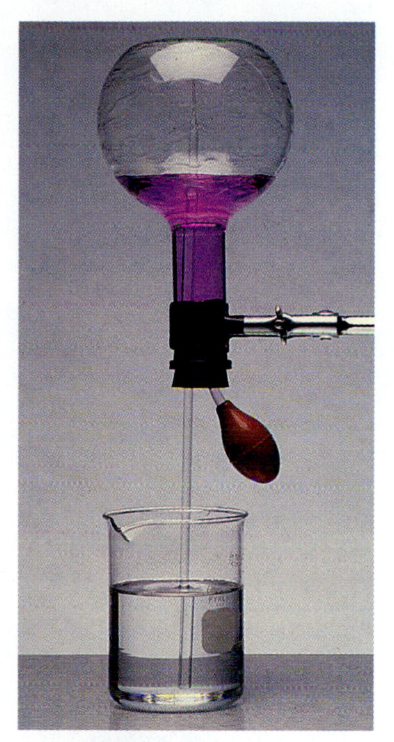

FIGURE 1.8

The "ammonia fountain." Initially the top flask holds only gaseous ammonia. Then the flask is inverted so that the fitted glass tube extends beneath the water surface in the beaker, which also has a drop of phenolphthalein indicator. The bulb is next used to squirt a small amount of water into the flask. As the gaseous ammonia dissolves in this water, a vacuum is created in the flask, which promptly draws more water from the beaker. The phenolphthalein makes the ammonia solution in the flask pink.

Basicity of Ammonia The most familiar reactions of ammonia involve its basicity, but recall that it is not a strong base, when compared with the OH^- ion in water, and has a small K_b.

$$NH_3(aq) + H_2O \rightleftharpoons NH_4^+(aq) + OH^-(aq) \quad K_b = 1.8 \times 10^{-5}\ (pK_b = 4.74)$$

A strong acid quantitatively converts ammonia to the ammonium ion. For example,

$$NH_3(aq) + HCl(aq) \xrightarrow{100\%} NH_4Cl(aq) + H_2O$$

or

$$NH_3(aq) + H_3O^+(aq) \longrightarrow NH_4^+(aq) + H_2O$$

Oxidation of Ammonia Ammonia is not considered to be highly flammable, but it can be made to burn in air to give nitrogen and water vapor.

$$4NH_3(g) + 3O_2(g) \rightleftharpoons 2N_2(g) + 6H_2O(g) \qquad K_p = 10^{228}\ (25\ °C)$$

The exceptionally high equilibrium constant for this reaction tells us that the forward reaction is strongly favored, thermodynamically. Nonetheless, ammonia is very stable in air; the forward reaction actually has several steps, and the rate-limiting step evidently has such a high energy of activation that it cannot occur at room temperature. However, the combustion is exothermic enough so that when gaseous ammonia is mixed with oxygen in the right proportion, an explosive combustion can be initiated.

In the presence of a platinum–rhodium catalyst at 750 to 900 °C, the oxidation of ammonia forms a species thermodynamically less favored than N_2, namely, nitrogen monoxide, NO (see Figure 1.9).

$$4NH_3(g) + 5O_2(g) \underset{750-900\ °C}{\overset{Pt/Rh}{\rightleftharpoons}} 4NO(g) + 6H_2O(g) \qquad K_p = 10^{168}\ (25\ °C)$$

This reaction is vital to the Ostwald synthesis of nitric acid from ammonia, as we will study soon.

Liquid Ammonia as a Solvent Liquid ammonia dissolves the Group IA and IIA metals, giving solutions with a blue color caused by ammoniated

Ammoniated species are indicated in equations by *(am)* after the formula, as in M^+ *(am)* and e^-*(am)*.

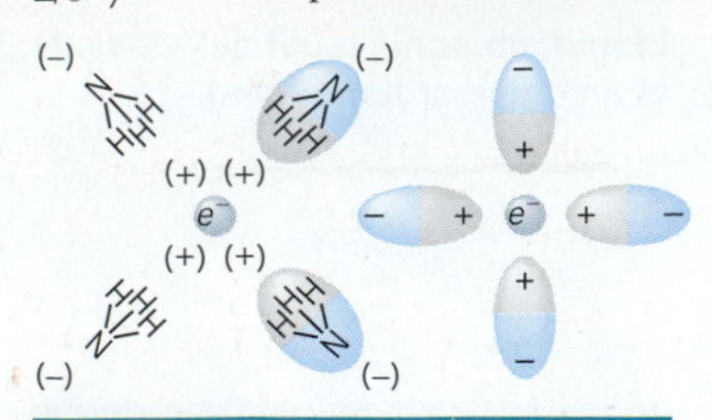

In the ammoniated electron system, molecules of NH_3 surround the electron.

FIGURE 1.9

Ammonia escaping from the aqueous ammonia at the bottom of this beaker reacts with oxygen at the surface of the catalyst, a gauze of platinum. Enough heat is generated to make the gauze glow. Colorless NO, initially produced, soon reacts with oxygen to give NO_2, which is responsible for the reddish-brown color.

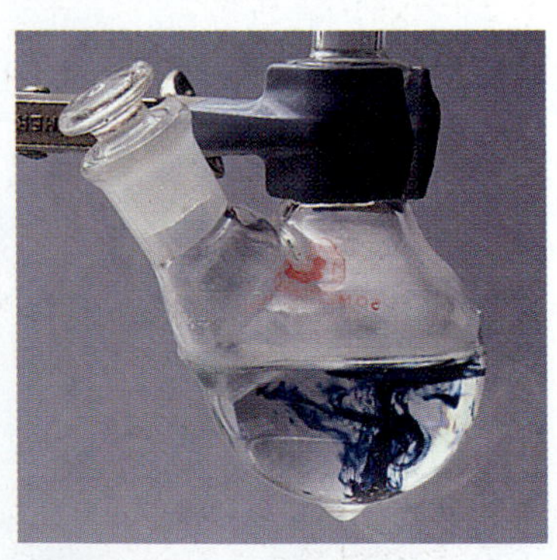

FIGURE 1.10

The liquid in the flask is liquid ammonia. The deep blue swirls are caused by ammoniated electrons that form as sodium dissolves in the ammonia.

electrons (see Figure 1.10). To be *ammoniated* means to be surrounded by molecules of the solvent, liquid ammonia. (Cations being ammoniated in liquid ammonia is like their being *hydrated* in water, namely, surrounded by solvent molecules.) Eventually, an irreversible reaction occurs between ammoniated electrons and liquid ammonia, so the end product is a liquid ammonia solution of the metal amide. For example, a solution of sodium in liquid ammonia changes as follows, and this is how sodium amide is manufactured.

$$2Na^+(am) + 2e^-(am) + 2NH_3(l) \longrightarrow H_2(g) + \underset{\text{sodium amide}}{2NaNH_2(am)}$$

Transition metal ions catalyze the reaction.

Acid–Base Chemistry in Liquid Ammonia Just as the hydroxide ion is the most powerful base that can exist in water, so the amide ion is the most powerful base that can exist in liquid ammonia. Similarly, the NH_4^+ ion in liquid ammonia is analogous to the H_3O^+ ion in liquid water—each cation is the strongest proton-donating species that can exist in its respective solvent. Thus, a whole range of acid–base chemistry exists for liquid ammonia involving NH_4^+ and NH_2^- as the key acidic and basic species. The important acid–base neutralization in liquid ammonia is

$$NH_4^+(am) + NH_2^-(am) \xrightarrow[NH_3(l)]{} 2NH_3(l)$$

Thus, we can titrate NH_4Cl with KNH_2 in liquid ammonia just as we can titrate HCl with KOH in water, and we can even use phenolphthalein as the indicator for both titrations.

Metal Amides Sodium and potassium amides are commercially available. Because the amide ion is a powerful base toward water, these metal amides react rapidly, quantitatively, and exothermically with water. The reaction with sodium amide, for example, is

$$NaNH_2(s) + H_2O \longrightarrow NH_3(aq) + NaOH(aq)$$

But the actual reaction is simply that of the amide ion.

$$NH_2^-(s) + H_2O \xrightarrow{100\%} NH_3(aq) + OH^-(aq)$$

Metal amides, therefore, must be stored in a dry atmosphere.

Hydrolysis of the Ammonium Ion The common inorganic ammonium salts are those with the anions Cl^-, NO_3^-, SO_4^{2-}, and the various anions from phosphoric acid. Generally, all ammonium salts dissolve in water and become fully dissociated.

The ammonium ion hydrolyzes, that is, it reacts somewhat with water in the following equilibrium to render slightly acidic any solution of a salt made of the ammonium ion and a nonhydrolyzing anion (like Cl^-).

$$NH_4^+(aq) + H_2O \rightleftharpoons NH_3(aq) + H_3O^+(aq)$$

Water is a weak proton acceptor, so the forward reaction is not favored. However, if a strong base, like OH^-, is added to an ammonium salt, the base is neutralized by NH_4^+ and ammonia is released. For example,

$$NH_4Cl(aq) + NaOH(aq) \longrightarrow NH_3(aq) + NaCl(aq) + H_2O$$

For NH_4^+, $K_a = 5.5 \times 10^{-10}$. The pH of 1 *M* NH_4Cl is about 4.7.

or

$$NH_4^+(aq) + OH^-(aq) \longrightarrow NH_3(aq) + H_2O$$

Decomposition of the Ammonium Ion Several ammonium salts decompose when heated strongly. When the anion of the ammonium salt is *not* an oxidizing agent, then one product is ammonia. Ammonium chloride, for example, sublimes without melting and decomposes at about 500 °C.

$$NH_4Cl(s) \xrightarrow{500\ ^\circ C} NH_3(g) + HCl(g)$$

When the anion of the ammonium salt is an oxidizing agent, like NO_3^- or $Cr_2O_7^{2-}$, thermal decomposition of the salt oxidizes its ammonium ion, and the products are (rapidly expanding) gases. These reactions are potentially very dangerous and should be attempted only by experienced chemists. When ammonium nitrate is heated, for example, the nitrate ion oxidizes the ammonium ion and dinitrogen monoxide forms.

$$\underset{\text{ammonium nitrate}}{NH_4NO_3(s)} \xrightarrow{\text{heat}} \underset{\text{dinitrogen monoxide}}{N_2O(g)} + 2H_2O(g)$$

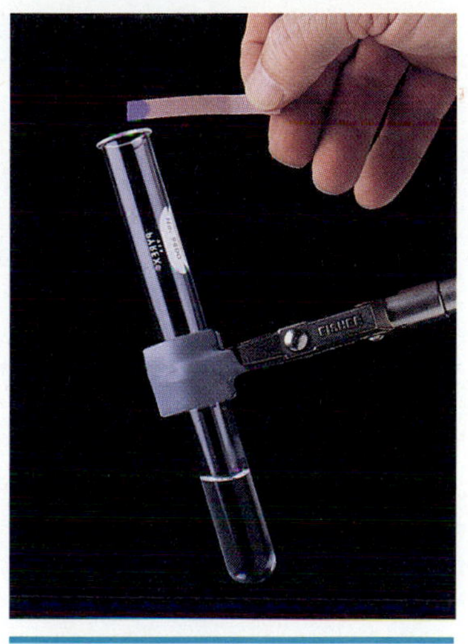

When base is added to a solution of an ammonium salt, NH_3 is generated. Some escapes to turn a moist litmus from red to blue.

The N of the ammonium ion in NH_4NO_3 is oxidized from the −3 to the +1 oxidation state, and the N of the nitrate ion is reduced from the +5 to the +1 state. When ammonium dichromate is heated, the N of its NH_4^+ ion is oxidized from its −3 to the 0 oxidation state by the dichromate ion, and Cr in $Cr_2O_7^{2-}$ is reduced from its +6 to its +3 state.

$$\underset{\text{ammonium dichromate}}{(NH_4)_2Cr_2O_7} \xrightarrow{\text{Heat}} N_2(g) + \underset{\text{chromium(III) oxide}}{Cr_2O_3(s)} + 4H_2O(g)$$

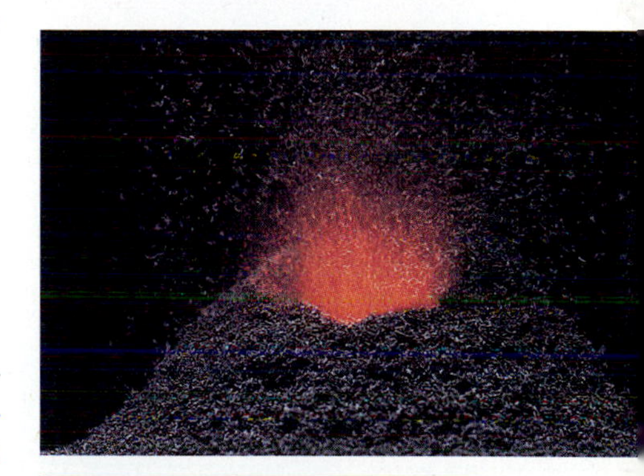

The ammonium dichromate "volcano." When heated, a mound of $(NH_4)_2Cr_2O_7$ decomposes dramatically as its gaseous products evolve.

Nitrides

The binary compounds between nitrogen in the −3 oxidation state and elements other than hydrogen are called **nitrides,** but the N has the same oxidation state as it does in ammonia (−3). Magnesium nitride, a typical metal nitride, is made by heating magnesium with nitrogen or with ammonia.

$$3Mg(s) + 2NH_3(g) \xrightarrow{900\ ^\circ C} \underset{\text{magnesium nitride}}{Mg_3N_2(s)} + 3H_2(g)$$

The nitrides of Group IA and IIA metals are saltlike compounds with high melting points, suggesting that they are ionic, the anion being the nitride ion, N^{3-}. Such metal nitrides react with water to give a metal hydroxide and ammonia, illustrated by the reaction of lithium nitride with water.

$$Li_3N(s) + 3H_2O \longrightarrow 3LiOH(aq) + NH_3(aq)$$

Nonmetal Nitrides The nitrides of nonmetallic elements, for example, BN, $(CN)_2$, S_4N_4, and P_3N_5, involve largely covalent bonds, but their properties differ widely. One form of boron nitride (BN) melts at 3000 °C and is very inert, but the nitride of carbon, cyanogen, $(CN)_2$, is a poisonous gas! The nitride of sulfur, S_4N_4, melts at 178 °C but may detonate when struck or heated too rapidly.

When compounds have high melting points we normally think that they are ionic, but not so with boron nitride. Instead, it is *macromolecular,* meaning that its molecules consist of thousands of covalently bound B and N atoms held in vast sheets, which are stacked one on top of the other (Figure 1.11). Thus, a high melting point can mean either ionic bonds or a macromolecular system involving covalent bonds. Obviously, BN is an empirical, not a molecular formula, whereas $(CN)_2$ and S_4N_4 are molecular formulas.

Transition Metal Nitrides Many transition metals form nitrides, including some that are almost as hard as diamond—for example, vanadium nitride, VN—and some that retain the appearance and electrical conductivity of a metal. We leave the nature of the bonds in these substances to other treatments, but we have learned here to be somewhat cautious in automatically assuming that a high melting point means that a compound must be ionic.

Other Important Compounds of Nitrogen in Its Reduced State Nitrogen occurs in the −2 oxidation state in *hydrazine,* NH_2NH_2 (often written as N_2H_4). It is a colorless, *toxic* liquid that boils at 113.5 °C and freezes at 2 °C, and it is a weaker base than ammonia. Yet, hydrazine is a strong enough base to form a family of salts of the *hydrazinium ion,* $N_2H_5^+$.

Base	K_b (25 °C)
N_2H_4	9.6×10^{-7}
NH_3	1.8×10^{-5}

Hydrazine is made commercially by the *Raschig process,* in which ammonia is oxidized by sodium hypochlorite.[2] The overall reaction is

$$2NH_3(aq) + NaOCl(aq) \longrightarrow N_2H_4(aq) + NaCl(aq) + H_2O$$

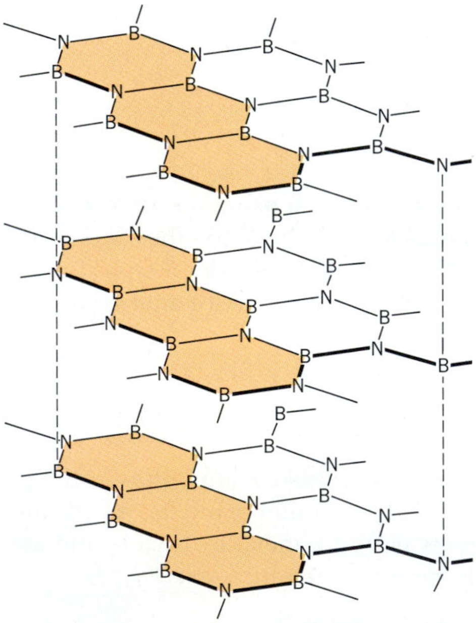

FIGURE 1.11

Boron nitride consists of stacks of sheets of macromolecules involving covalent bonds between B and N.

[2] The "chlorine" bleaches used in home laundries contain the hypochlorite ion, OCl^-, so they must never be mixed with household ammonia. NH_3 and OCl^- react to give hydrazine and some chloramine (NH_2Cl), both of which are toxic. Fortunately, plumbing made of copper pipe releases traces of Cu^{2+} ion, which catalyze the decomposition of hydrazine. Thus, as laundry products drain away, any hydrazine that may have formed is decomposed.

Hydrazine is a strong reducing agent, as indicated in the following half-cell reaction, which is just one mode by which hydrazine can function. The large *negative* value of $E°$ indicates that the reverse reaction, which would supply electrons to some acceptor (and thus reduce it) is favored.

$$N_2(g) + 4H_2O + 4e^- \rightleftharpoons N_2H_4(aq) + 4OH^-(aq) \qquad E° = -1.16\ V$$

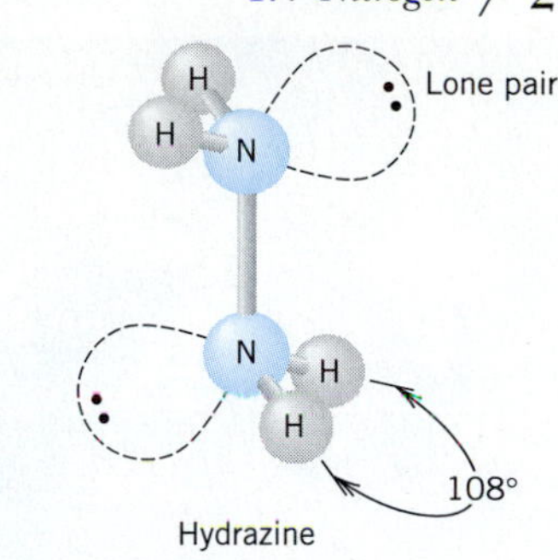

The combustion of dimethylhydrazine, $(CH_3)_2NNH_2$, and oxygen was used to propel United States astronauts to the moon and back.

Liquid hydrazine and some of its organic derivatives have been used as rocket fuels. When mixed with hydrogen peroxide or with oxygen from liquid oxygen tanks, hydrazine burns violently to produce rapidly expanding gases that give thrust to the rocket. The reaction is very exothermic. For example, the oxidation of N_2H_4 by O_2 is

$$N_2H_4(l) + O_2(g) \longrightarrow N_2(g) + 2H_2O(l) \qquad \Delta H° = -621.5\ kJ$$

Nitrogen is in the -1 oxidation state in *hydroxylamine,* $HONH_2$, a white solid (MP 33 °C). We can think of the $HONH_2$ molecule as NH_3 in which one H has been replaced by OH, or as H_2O in which one H has been replaced by NH_2. Hydroxylamine is basic; $K_b = 6.6 \times 10^{-9}$ (25 °C).

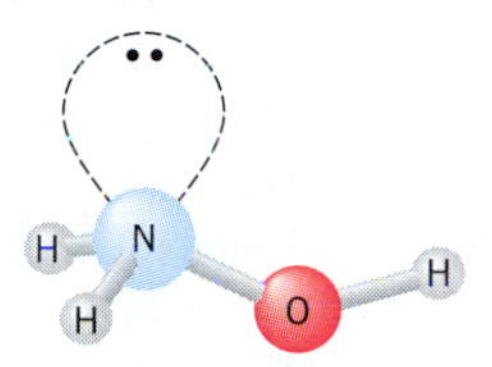

Nitrogen is in the unusual oxidation state of $-1/3$ in *hydrazoic acid,* HN_3, a weak acid (K_a, 1.8×10^{-5}; pK_a, 4.74, 25 °C). When pure, hydrazoic acid is a colorless liquid and extremely susceptible to explosion. The azide ion is linear and symmetrical. In valence bond theory it is described as the hybrid of the following resonance structures.

$$[:N\equiv N-\ddot{\underset{..}{N}}:]^- \longleftrightarrow [:\ddot{N}=N=\ddot{N}:]^- \longleftrightarrow [:\ddot{\underset{..}{N}}-N\equiv N:]^-$$

As these structures suggest, the two nitrogen–nitrogen bonds should be, and are, identical in length, namely, 116 pm. The azide ion in aqueous solution behaves something like a halide ion and is sometimes called a *pseudohalide ion.*

Several salts called *azides* are known. Those of heavy metals, like lead azide, explode when sharply struck, so they are used to make detonator caps, which are devices for setting off the explosions of other materials, like gunpowder. The azides of the Group IA metals do not explode so easily, and sodium azide, NaN_3, is used in vehicle airbags. When activated by an electric spark, sodium azide *very* rapidly decomposes to give (rapidly expanding) nitrogen gas.

$$2NaN_3(s) \longrightarrow 2Na(l) + 3N_2(g)$$

The other product, sodium, forms in the molten state, but the airbag system also contains iron(III) oxide, which combines exothermically with the sodium as it forms to change it to a less hazardous sodium oxide. The sudden appearance of nitrogen gas expands the airbag, but the bag is porous so it goes limp quickly enough to release its pressure on the vehicle occupant.

$$6Na(l) + Fe_2O_3(s) \longrightarrow 3Na_2O(s) + 2Fe(s)$$

Oxoacids and Oxides and of Nitrogen

Nitric Acid, Dinitrogen Pentoxide, and the Nitrates Nitrogen is in its highest oxidation state, $+5$, in nitric acid, dinitrogen pentoxide (nitric anhydride), and the nitrates.

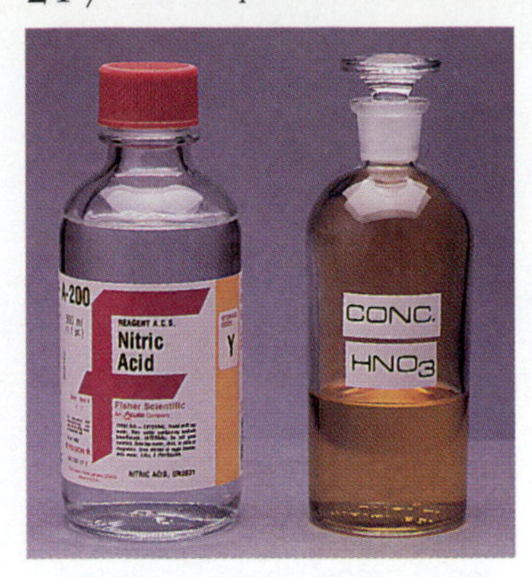

Concentrated nitric acid. On the left is a fresh bottle in which the acid is still colorless. On the right is an aged sample, now colored by a trace of NO_2.

The violent decomposition of nitroglycerin produces 35 mol of hot gases from just 4 mol of a solid, which causes a sudden expansion.

$$4C_3H_5N_3O_9 \longrightarrow 6N_2 + 12CO + 7O_2 + 10H_2O \text{ (as steam)}$$

The symbol (v/v) means percentage by volume.

Dynamite sticks are widely used in mining.

nitric acid — dinitrogen pentoxide or nitric anhydride (gas-phase structure) — nitrate ion

Pure nitric acid is known only in its solid state. In the liquid or gaseous states, it spontaneously decomposes to nitrogen dioxide and oxygen.

$$4HNO_3 \longrightarrow 4NO_2 + 2H_2O + O_2$$

Sunlight accelerates this reaction, which also slowly happens in aqueous nitric acid. (Because NO_2 is a reddish-brown gas, aged nitric acid solutions generally have a yellow to reddish color.)

Ostwald Process for Nitric Acid Nitric acid is manufactured from ammonia by the **Ostwald process,** developed in 1902 by Wilhelm Ostwald (Germany; 1853–1932; Nobel Prize, 1909). Historians believe that Ostwald's method for making nitric acid significantly prolonged the First World War, because nitric acid or its nitrate salts are needed to make modern explosives, like TNT (trinitrotoluene) and nitroglycerin for gunpowder. An Allied naval blockade had prevented Germany from importing nitrates for this purpose from Chile. Germany's ability to make its own nitrates by the Ostwald process, however, freed it from these supplies. Alfred Nobel, the inventor of dynamite, discovered a relatively safe way to handle nitroglycerin, and his fortune now funds the Nobel Prizes.

In modern operations of the Ostwald process, a mixture of air with 10% (v/v) NH_3, heated to 850 °C and pressurized to 5 atm, flows through a series of gauzes made of an alloy of platinum and rhodium, which serve as the catalyst. The first reaction is the oxidation of ammonia to nitrogen monoxide, which we discussed earlier.

$$4NH_3(g) + 5O_2(g) \xrightarrow[750-900\ ^\circ C]{Pt/Rh} 4NO(g) + 6H_2O(g) \qquad \Delta H^\circ = -1170\ kJ$$

As you can see, this step is very exothermic and the flowing gas stream has to be cooled somewhat. Thus, more air is let in, and the nitrogen monoxide reacts with oxygen to give nitrogen dioxide.

$$2NO(g) + O_2(g) \longrightarrow 2NO_2(g)$$

The flowing gases are then passed into a spray of water in special absorbing towers where newly formed NO_2 reacts with water to give 60% nitric acid.

$$3NO_2(g) + H_2O \longrightarrow 2HNO_3(aq) + NO(g)$$

Unchanged components of the initial air plus the NO produced in this last step leave the absorbing towers and are recycled. The overall change in the Ostwald process can be written as follows.

$$NH_3(g) + 2O_2(g) \longrightarrow HNO_3(aq) + H_2O$$

The 60% nitric acid produced in the absorbing towers is concentrated to a maximum of 68.5% HNO_3 by distillation. At this concentration, the solution boils at a constant temperature and cannot be made more concentrated by further distillation. Special methods are needed to obtain a more concentrated nitric acid.

Commercial "concentrated HNO_3" has a concentration of 16 *M*.

Nitric acid is one of the top 20 chemicals manufactured worldwide. By far its largest use (80%) is to make ammonium nitrate, NH_4NO_3, for fertilizers.

$$NH_3 + HNO_3 \longrightarrow NH_4NO_3$$

Some ammonium nitrate is used to make dynamite, which today is a mixture of ammonium nitrate, nitroglycerin, sodium nitrate, wood pulp, and a trace of calcium carbonate.

Chemical Properties of Nitric Acid Nitric acid, although a strong monoprotic acid, is used more as an *oxidizing agent* than as an acid. Depending on the conditions, the +5 state of N in the nitrate ion can be reduced to the +4 state (in NO_2), the +3 state (in HNO_2), or the +2 state (in NO). A typical oxidation by nitric acid is its reaction with copper in which the products depend on the concentration of the acid. *Concentrated* nitric acid favors the formation of nitrogen dioxide, NO_2.

$$Cu(s) + 4HNO_3(aq) \xrightarrow{\text{In concentrated } HNO_3} Cu(NO_3)_2(aq) + 2NO_2(g) + 2H_2O$$

Dilute nitric acid favors the formation of nitrogen monoxide, NO.

$$3Cu(s) + 8HNO_3(aq) \xrightarrow{\text{In dilute } HNO_3} 3Cu(NO_3)_2(aq) + 2NO(g) + 4H_2O$$

Dinitrogen Pentoxide (Nitric Anhydride) The acidic anhydride of nitric acid, dinitrogen pentoxide or N_2O_5, is a white solid made by letting P_4O_{10} remove water from concentrated nitric acid by the following reaction.

P_4O_{10} is a powerful dehydrating agent.

$$\underset{}{4HNO_3(aq)} + \underset{\substack{\text{tetraphosphorus} \\ \text{decaoxide}}}{P_4O_{10}(s)} \xrightarrow{-10\ °C} 2N_2O_5(s) + \underset{\substack{\text{metaphosphoric} \\ \text{acid}}}{4HPO_3(l)}$$

In the *solid* state, N_2O_5 is an ionic compound, existing as a combination of the nitronium ion, NO_2^+, and the nitrate ion, in other words, as nitronium nitrate ($NO_2^+NO_3^-$). In the *gas* phase, N_2O_5 is a molecular compound with the structure shown earlier.

Solid dinitrogen pentoxide sublimes at 32.4 °C, but it must be handled very cautiously, because it can explode.

Nitrate Salts Nearly all metals have nitrate salts, but they vary widely in thermal stability. Those of Groups IA and IIA are particularly stable, but at high enough temperatures they decompose to the corresponding nitrites and oxygen. When the nitrite is itself thermally unstable, the decomposition continues until a metal oxide forms. Note, for example, the different ways that sodium and potassium nitrates decompose, one to a thermally stable nitrite and the other to an oxide.

Nitrate	Melting Point (°C)
$LiNO_3$	255
$NaNO_3$	303
KNO_3	310
$CsNO_3$	414

$$2NaNO_3(s) \xrightarrow{>500\ °C} 2NaNO_2(s) + O_2(g)$$

$$4KNO_3(s) \xrightarrow{>500\ °C} 2K_2O(s) + 2N_2(g) + 5O_2(g)$$

Either a detonator or a high temperature causes ammonium nitrate to explode violently, particularly when some easily oxidized impurity is present. All the products form as (rapidly expanding) gases.

A major part of Texas City, Texas, was destroyed in April 1947 from an explosion of ammonium nitrate being loaded into a ship. Nearly 600 lives were lost. NH_4NO_3 was also the explosive used in the bombing of the Federal Building in Oklahoma City in 1995.

$$2NH_4NO_3(s) \xrightarrow{>300\ ^\circ C} 2N_2(g) + O_2(g) + 4H_2O(g)$$

Under milder conditions, ammonium nitrate decomposes to dinitrogen monoxide, and the reaction is used to prepare this anesthetic gas.

$$NH_4NO_3(s) \xrightarrow{200\text{–}260\ ^\circ C} N_2O(g) + 2H_2O(g)$$

Nitrous Acid, Dinitrogen Trioxide, and the Nitrites Nitrogen is in its +3 oxidation state in nitrous acid, dinitrogen trioxide (nitrous anhydride), and the nitrite salts.

nitrous acid | dinitrogen trioxide or nitrous anhydride | nitrite ion

Nitrous acid, HNO_2, is a weak, thermally unstable acid known only in solution, never as a pure compound. When aqueous nitrous acid is needed (and the presence of other ions is unimportant), chemists simply mix cold hydrochloric acid with an equimolar amount of sodium nitrite. The nitrite ion, the conjugate base of nitrous acid, is a good Brønsted base and accepts a proton to form HNO_2.

$$NaNO_2(aq) + HCl(aq) \xrightarrow{0\ ^\circ C} HNO_2(aq) + NaCl(aq)$$

Nitrous acid (K_a, 7.1×10^{-4}) is a slightly stronger acid than acetic acid (K_a, 1.8×10^{-5}). As we said, it is thermally unstable and decomposes, when heated, largely as follows, with one NO_2^- ion being oxidized to NO_3^- and two NO_2^- ions being reduced to 2NO.

$$3HNO_2(aq) \xrightarrow{\text{Heat}} HNO_3(aq) + 2NO(g) + H_2O$$

Nitrous acid reacts either as an oxidizing agent or a reducing agent, depending on the other reactant. For example, HNO_2 is an oxidizing agent toward the iodide ion, a good reducing agent, and is reduced to nitrogen monoxide, NO.

$$2I^-(aq) + 2HNO_2(aq) + 2H^+(aq) \longrightarrow I_2(s) + 2NO(g) + 2H_2O$$

Nitrous acid is a reducing agent toward the permanganate ion, a powerful oxidizing agent, and HNO_2 is oxidized to the nitrate ion.

$$2MnO_4^-(aq) + 5HNO_2(aq) + H^+ \longrightarrow 2Mn^{2+}(aq) + 5NO_3^-(aq) + 3H_2O$$

Dinitrogen trioxide, a deep blue liquid, is the acid anhydride of nitrous acid only in a formal sense. It is not made by removing water from nitrous acid but simply by combining NO and NO_2 in a 1:1 mole ratio at $-30\ ^\circ C$. At this low temperature, the forward reaction is favored:

$$NO(g) + NO_2(g) \rightleftharpoons N_2O_3(l)$$

As N_2O_3 is warmed above $-30\ ^\circ C$, the above equilibrium increasingly shifts to the left in favor of nitrogen monoxide and nitrogen dioxide.

Dinitrogen trioxide reacts with water to form nitrous acid.

$$N_2O_3(g) + H_2O \longrightarrow 2HNO_2(aq)$$

In fact, if the same mixture of gases that gives N_2O_3, namely, $NO + NO_2$, is bubbled into cold water, nitrous acid forms directly.

The most common *nitrite salt* is sodium nitrite, prepared industrially by bubbling an equimolar mixture of NO and NO_2 into aqueous sodium hydroxide and then evaporating the water.

Sodium nitrite is thermally stable, melting without decomposition at 284 °C.

$$2NaOH(aq) + NO(g) + NO_2(g) \longrightarrow 2NaNO_2(aq) + H_2O$$

Sodium nitrite is used as a food additive in meat products like corned beef, ham, and bacon, where it inhibits the growth of microorganisms that cause food poisoning. (It also maintains the reddish color of these products.) Unfortunately, the nitrite ion also affects the hemoglobin in blood, so the U.S. Food and Drug Administration allows only very small traces of sodium nitrite in meat products and none at all in infants' food.

Other Oxides of Nitrogen The oxides of nitrogen (Table 1.5) are compounds with very different reactivities. We have already studied two of them, N_2O_5 and N_2O_3. We'll next look briefly at the rest.

Dinitrogen monoxide (nitrous oxide), N_2O, widely used as a safe, nonexplosive anesthetic ("laughing gas"), has nitrogen in its +1 oxidation state. We noted earlier that N_2O is made by carefully heating ammonium nitrate. At room temperature, N_2O is quite unreactive toward most substances, including alkali metals, the halogens, and even ozone. The unreactivity makes N_2O useful as a propellant in aerosol cans.

N_2O is thermodynamically unstable with respect to its elements, as indicated by its *positive* standard heat of formation ($\Delta H_f^\circ = +81.5$ kJ mol^{-1}) and its *positive* standard free energy of formation ($\Delta G_f^\circ = +104$ kJ mol^{-1}). The *rate* of its decomposition at room temperature, however, is so slow that there is no problem in storing it, although when heated sufficiently, N_2O decomposes to nitrogen and oxygen.

$$2N_2O(g) \longrightarrow 2N_2(g) + O_2(g)$$

TABLE 1.5 Oxides of Nitrogen

Oxidation State of Nitrogen	Formula	Name	Color	Boiling Point (°C)	Melting Point (°C)
+1	N_2O	Dinitrogen monoxide (nitrous oxide)	None	−89	−91
+2	NO	Nitrogen monoxide (nitric oxide)	None	−152	−164
+3	N_2O_3	Dinitrogen trioxide (nitrous anhydride)	Red brown	Decomposes	−102
+4	NO_2	Nitrogen dioxide[a]	Red brown	−11[b]	
+4	N_2O_4	Dinitrogen tetroxide	None		−21
+5	N_2O_5	Dinitrogen pentoxide (nitric anhydride)	None	Decomposes	32

[a] NO_2 and N_2O_4 exist in the presence of each other in both the liquid and gaseous states. At 25 °C and 1 atm, N_2O_4 is favored; its partial pressure is about 540 torr and that of NO_2 is about 220 torr. In the solid state, the form is N_2O_4.

[b] The temperature at which the *mixture's* vapor pressure equals 1 atm is −11 °C.

Notice that 3 mol of gaseous products form from 2 mol of the gaseous reactant. Therefore, at a high temperature in a closed container, like inside a vehicle engine cylinder, the reaction causes an increase in pressure. Some drivers of racing cars, therefore, inject N_2O into their fuel lines to give more power to the engine. Moreover, the reaction is exothermic, and the heat of decomposition adds to the heat of combustion of the fuel to give even more takeoff power. In addition, oxygen is a product, and it promptly enters into the combustion of the fuel, providing a still further boost.

The electron configuration of the NO molecule in molecular orbital (MO) theory is similar to that of O_2 (Table 8.1), but NO has one more electron. This is unpaired, so NO is paramagnetic.

Nitrogen monoxide (nitric oxide), NO, with N in the +2 oxidation state, can be made by the reaction of dilute nitric acid with a relatively unreactive metal, like copper, as we studied earlier. NO also forms from the N_2 and O_2 in air when fuels are burned in vehicles.

$$N_2(g) + O_2(g) \rightleftharpoons 2NO(g) \qquad K_p = 4.1 \times 10^{-31}\ (25\ ^\circ C)$$

The extremely small equilibrium constant means that the *reactants* are strongly favored at 25 °C. Thus, essentially no forward reaction occurs in air at ordinary temperatures (which is nice, because otherwise little if any oxygen would be left for us to breathe). At the high temperature in an engine cylinder, however, the equilibrium constant is somewhat higher, so some NO does form. As the exhaust gases cool, the NO does not immediately change back to N_2 and O_2, because the energy of activation for this change is high enough to make the process slow. Thus, NO is a by-product of the combustion of fuels, and it is a significant contributor to the formation of other air pollutants, notably, NO_2.

Whenever NO forms and mixes with air at moderate to low temperatures, it combines quickly with oxygen to form nitrogen dioxide, NO_2.

$$2NO(g) + O_2(g) \longrightarrow 2NO_2(g)$$

Thus, when hot auto exhaust gases with traces of NO reach the outside air and are cooled, the NO is oxidized to NO_2. This gas gives the characteristic reddish-brown color to smog (Figure 1.12).

NO_2 is paramagnetic because each molecule has an unpaired electron. N_2O_4, in which all electrons are paired, is diamagnetic.

In *nitrogen dioxide,* NO_2, and *dinitrogen tetroxide,* N_2O_4, nitrogen is in the +4 oxidation state. These two gases occur together in an equilibrium.

FIGURE 1.12

(*Left*) The Los Angeles skyline on a clear day. (*Right*) The same view during smog. The reddish color is caused by NO_2.

$$2NO_2 \rightleftharpoons N_2O_4$$

nitrogen dioxide (reddish brown) — dinitrogen tetroxide (colorless)

The position of this equilibrium and the color of the system vary with temperature (Figure 1.13). Pure, solid N_2O_4 is present below −21 °C, the melting point of the system, so there is no color. At −21 °C (the melting point of N_2O_4), there is 0.01% NO_2, enough to make the system pale yellow. The mixture boils at 21 °C, and now it contains 0.1% NO_2, enough to impart a deep reddish-brown color. Above 140 °C, the system is 100% NO_2 and an even darker reddish brown.

We have already noted that NO_2 forms when NO mixes with oxygen at ordinary temperatures. The NO_2 is the dangerous member of this pair. Besides contributing the disagreeable color to smog, it is itself poisonous, and to make matters worse, it reacts with rainwater to generate a dilute solution of nitric acid and nitrous acid.

$$2NO_2(g) + H_2O \longrightarrow HNO_3(aq) + HNO_2(aq)$$

This reaction is therefore one source of the acid in *acid rain* (see *Chemicals in Use* 10).

FIGURE 1.13

The position of the equilibrium, $2NO_2 \rightleftharpoons N_2O_4$, depends on the temperature. (*Left*) At the warmer temperature, the red-brown color of NO_2 is evident. (*Right*) At the colder temperature, the color is much less pronounced because the equilibrium has shifted to the right, in favor of the colorless N_2O_4.

1.5 CARBON

The **carbon family,** Group IVA, consists of carbon, silicon, germanium, tin, and lead (see Table 1.6). Carbon is the chief constituent of coal and forms the backbones of the hydrocarbon molecules in oil and natural gas. Carbon also occurs widely in carbonate rocks, such as limestone, dolomite, and marble.

TABLE 1.6 The Carbon Family

Element	Symbol	Melting Point (°C)	Boiling Point (°C)	Important Uses and Types of Compounds
Carbon	C	3800[a]	—	Diamond, graphite, coke, activated charcoal; carbon dioxide, carbonates, cyanides, carbides, organic compounds
Silicon	Si	1420	3280	Transistors, computer chips, silicates (e.g., quartz sand), silicone oils and greases
Germanium	Ge	959	2700	Transistors, phosphors, special alloys
Tin	Sn	232	2275	Tin plate ("tin cans") Solder (Sn/Pb alloy) Bronze (Sn/Cu alloy) Pewter (Sn/Sb/Cu alloy) Type metal (Sn/Sb/Pb alloy)
Lead	Pb	327	1750	Batteries, lead-based pigments, alloys (e.g., solder, type metal), gasoline additive (lead tetraethyl)

[a] Graphite under pressure to suppress sublimation.

TABLE 1.7 Hydrides and Tetrachlorides of the Group IVA Elements

	Boiling Point (°C)	
Element	MH_4	MCl_4
Carbon	−161.4	76.7
Silicon	−111.8	59.0
Germanium	−88.1	83.1
Tin	−55.5	114
Lead	Unknown	*a*

[a] Unstable above 0 °C.

All of the Group IVA elements form oxides of the general formula MO_2 (where M is the particular member of the family). All form tetrahalides, MX_4, but only PbF_4 of the lead halides is stable at room temperature. Except for lead, the Group IVA elements form hydrides of the formula MH_4 (see Table 1.7). Carbon, alone among all of the elements, is able to form chains of thousands of atoms and rings from three carbon atoms in size up to any size.

The centerpiece of this necklace is the famous Hope diamond, which has a brilliant blue color, unusual for a diamond.

The Chief Allotropes of Carbon

The two most common allotropes of carbon are graphite and diamond. In graphite, the carbon atoms occur in parallel sheets (see Figure 1.14*a*). Each sheet is made of extended networks of carbon hexagons within which occur vast networks of delocalized pi electrons. They permit graphite to conduct electricity, which is unusual for a nonmetal. The sheets are held to each other by relatively weak forces, and it is easy for planes to slide by each other. This is why powdered graphite is a good dry lubricant, and you may have used it to lubricate locks.

The carbon atoms in diamond (Figure 1.14*b*) are joined exclusively by sigma bonds that project tetrahedrally from each carbon. Gem diamonds have, of course, intrigued and attracted all who have ever looked upon them.

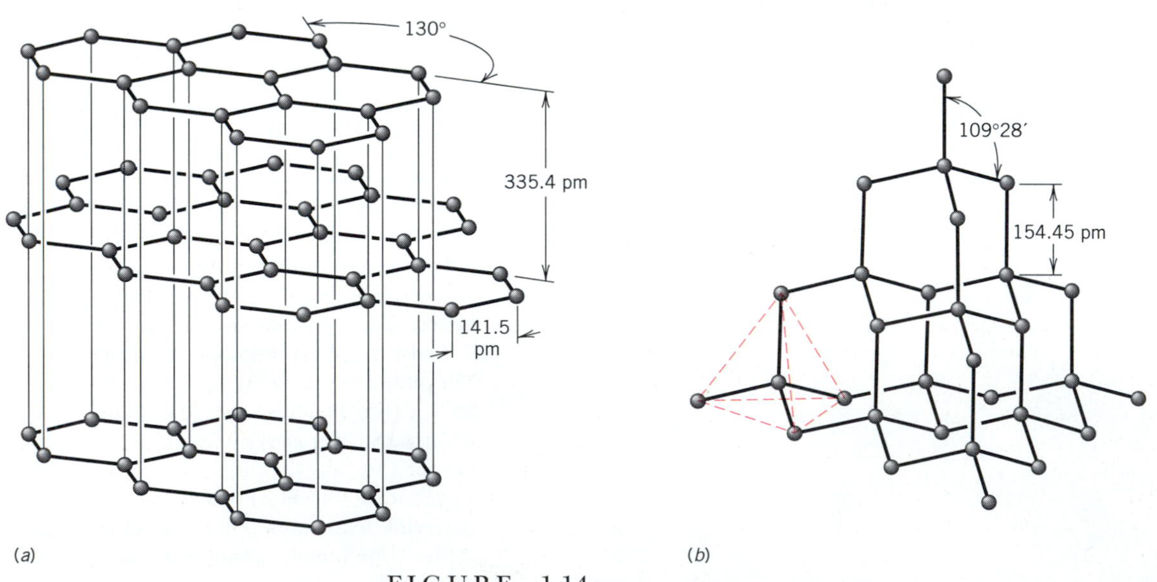

FIGURE 1.14

Two allotropes of carbon. (*a*) Graphite structure. (*b*) Diamond structure.

Thermodynamically, diamond is slightly *less* stable than graphite at ordinary pressures. But at ultrahigh pressures and temperatures, graphite can be forced into the less stable form, diamond (see *Chemicals in Use* 11).

For graphite, $\Delta H_{combustion}$ = − 393.51 kJ mol^{-1}.
For diamond, $\Delta H_{combustion}$ = − 395.41 kJ mol^{-1}.

Charcoal is mostly graphite in structure, but its preparation by strongly heating wood in the absence of air gives charcoal a very porous open structure with a huge surface area per gram. *Activated charcoal,* a very finely powdered form of charcoal that has been heated to drive off adsorbed molecules of other substances, has as much surface area as 10^3 m^2/g. Its ability to adsorb polar molecules to its surface accounts for its use to remove pollutants from water and air.

A baseball infield has an area of roughly 103 m^2 (10.8 × 10^3 ft^2).

Coke resembles charcoal but is made by heating soft coal under conditions in which it does not burn. Coke is widely used in metallurgy to reduce oxide ores to their metals (see Chapter 3 of this supplement).

Carbon black, still another form of finely divided carbon, is made by the combustion of a hydrocarbon in a limited amount of oxygen and is widely used to make vehicle tires and printing inks.

$$CH_4(g) + O_2(g) \longrightarrow \underset{\text{carbon black}}{C(s)} + 2H_2O(g)$$

The *fullerenes,* an exciting recent development in carbon chemistry, appeared in the late 1980s with the discovery of a soccerball-shaped molecule, C_{60}, in the products of the combustion of certain fuels (Figure 1.15). C_{60} is now made in quantity from graphite in arc furnaces or by laser technology. Because the structure of C_{60} resembles the geodesic dome pioneered by architect Buckminster Fuller, its discoverers named C_{60} "buckminsterfullerene." Today, the growing family of this and similar materials is called the *fullerenes.* Variations of C_{60}, like C_{70}, are also known, and derivatives, such as K_3C_{60}, have shown promise as superconducting materials.

C_{60} molecules have also been dubbed "buckyballs."

Oxides of Carbon

Carbon forms several oxides, but only CO and CO_2 are particularly stable.

Carbon Monoxide Carbon monoxide is used as a fuel, a reducing agent, and as a ligand in coordination compounds. Pure carbon monoxide is made

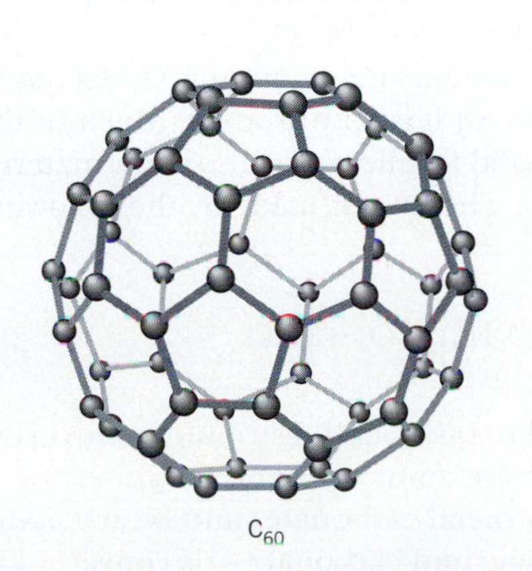

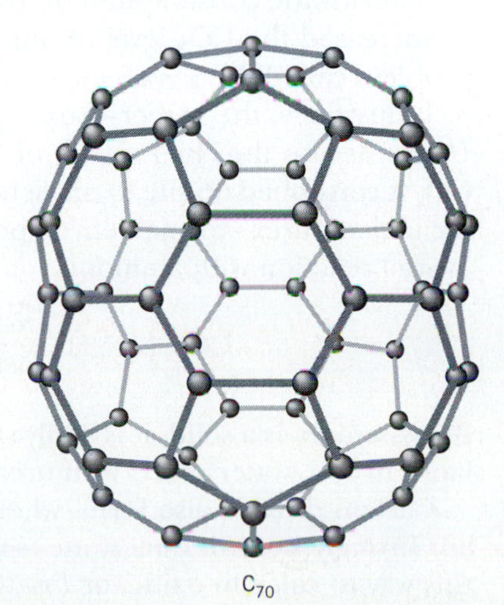

FIGURE 1.15
Two members of the fullerene family, C_{60} and C_{70}. C_{60} has a cage structure resembling a soccer ball and the C_{70} molecule is a larger version pinched at its "waist." In C_{60}, there are 60 vertices and 32 faces made up of 20 hexagons and 12 pentagons. Each carbon atom, being sp^2 hybridized, contributes an unhybridized $2p$ orbital to a huge, overlapping aromatic network, a pi electron cloud both inside and outside the structure.

on a small scale by heating formic acid with concentrated H_2SO_4, a strong dehydrating agent.

$$HCHO_2 \xrightarrow[\text{Heat}]{H_2SO_4\text{ (concd)}} CO(g) + H_2O$$

Tobacco smoke at the cigarette tip is about 2 to 5% carbon monoxide.

CO is always produced when any carbon-containing fuel burns incompletely. Thus, vehicle exhaust contains some, so CO is another component of smog.

Carbon monoxide burns in a sufficient supply of oxygen to carbon dioxide. In fact, CO is used industrially as a fuel whenever a chemical operation makes it as a by-product. Thus, the chemical energy of CO is not wasted, and burning CO to CO_2 disposes of a dangerous poison.

Because CO can be sent cheaply in pipelines, it has advantages over coke as a fuel. Coke, therefore, is sometimes converted by the action of superheated steam to a combustible, gaseous mixture of CO and H_2, called *water gas.*

$$C(s) + H_2O(g) \xrightarrow{1000\text{ °C}} CO(g) + H_2(g)$$

Carbon monoxide is also used as a reducing agent in the metallurgical industry to convert metal oxides to metals. For example,

$$FeO + CO \xrightarrow{\text{Heat}} Fe + CO_2$$

$$CuO + CO \xrightarrow{\text{Heat}} Cu + CO_2$$

CO binds about 200 times more strongly than O_2 to hemoglobin.

Carbon monoxide is an important ligand in the chemistry of coordination compounds, those that contain complex ions (see Section 19.5). It is this ligand property of CO that makes it a poisonous gas, because it binds to the Fe^{2+} ion of hemoglobin and blocks the ability of this oxygen carrier in blood to take up and transport oxygen.

Carbon Dioxide and Carbonic Acid Carbon dioxide is one of the end products of the complete combustion of organic compounds, including the fossil fuels. Octane, for example, burns as follows.

$$2C_8H_{18} + 25O_2 \longrightarrow 16CO_2 + 18H_2O$$

The worldwide consumption of fossil fuels during the last 100 years has steadily increased the CO_2 level of our atmosphere and contributed to a growing problem called the *greenhouse effect* (see *Chemicals in Use* 4).

Industrially, the Haber–Bosch process for making ammonia (*Chemicals in Use* 12) is also the chief source of CO_2, where it is a by-product. Much of the CO_2 is consumed on site to make urea, a solid fertilizer and also a raw material for making urea–formaldehyde polymers. The urea is made by the following overall reaction with ammonia.

Urea is the principal form in which nitrogen is excreted by mammals.

$$CO_2 + 2NH_3 \xrightarrow[185\text{ °C}]{200\text{ atm}} \underset{\text{urea}}{(NH_2)_2CO} + H_2O$$

Because urea is a solid, it is easily shipped to farms and distributed onto cropland. In soil, water reacts with urea to release ammonia (and CO_2).

Carbon dioxide also forms when many metal carbonates and bicarbonates are strongly heated. Limestone—mostly calcium carbonate—is converted in this way to calcium oxide, or *lime,* an important commercial material.

$$CaCO_3(s) \xrightarrow{\text{Heat}} \underset{\text{lime}}{CaO(s)} + CO_2(g)$$

Some common names for CaO are *lime, burnt lime, quick lime,* and *calx.* $Ca(OH)_2$ is called *hydrated lime* or *slaked lime.*

The reaction is carried out, however, to obtain lime, not CO_2. Between 16 and 18 million tons of lime are made annually in the United States. Lime is used mostly as an inexpensive base by the metallurgical, chemical, and wastewater treatment industries. Scrubbing SO_2 from smokestack gases, for example, is a growing use of lime. The smokestack gases pass over the lime, and the SO_2 reacts and thus is removed from the gases as solid $CaSO_3$.

Steel making uses lime as a fluxing agent to remove impurities in molten iron.

$$CaO + SO_2 \longrightarrow CaSO_3$$

The thermal decomposition of limestone also occurs in making *cement.* When finely ground limestone, sand, and clay are strongly heated together in rotating kilns, the product is transformed into *portland cement,* a mixture of various calcium silicates, such as Ca_2SiO_4 and Ca_3SiO_5, and calcium aluminates, like $Ca_3Al_2O_6$. When water and sand are added to portland cement, an unusually strong, stonelike solid called *concrete* forms that features large networks of —Si—O—Si—O—Si— systems.

The largest use of carbon dioxide—about 50%—is not as a chemical but as a refrigerant. CO_2 becomes a solid below −78.5 °C, and solid CO_2 is popularly known as "dry ice." Its triple point is above atmospheric pressure, so CO_2 cannot exist as a liquid except under pressure. It sublimes at ordinary pressures, but, if kept under 5.2 atm, CO_2 can be melted to a liquid at −56 °C. Liquid CO_2 is used as a solvent to extract caffeine from coffee to make "decaffeinated coffee." The advantage of this solvent is that it leaves no harmful residues in the coffee.

A familiar use of CO_2 is in a fire extinguisher. A blanket of the more dense CO_2, replacing the less dense air, shuts oxygen away from the burning material, and the fire goes out. CO_2 can be used on any kind of common fire—wood, paper, gasoline, or electrical.

About 25% of CO_2 production goes to the carbonation of beverages. A saturated solution of CO_2 in water at 760 torr and 25 °C is 0.033 *M* in dissolved CO_2—just as found in an opened bottle of club soda. Most of the CO_2 molecules exist in water simply as hydrates but a fraction are present as the unstable acid, carbonic acid, H_2CO_3.

$$CO_2(aq) + H_2O \rightleftharpoons H_2CO_3(aq)$$

The equilibrium strongly favors hydrated CO_2 molecules, the ratio $[CO_2(aq)]/[H_2CO_2(aq)]$ in saturated aqueous CO_2 being about 600:1. However, when it comes to the ability of aqueous CO_2 to neutralize base, all dissolved CO_2 is available, whatever the form, because CO_2 can react directly with aqueous hydroxide ion.

$$CO_2(aq) + OH^-(aq) \longrightarrow HCO_3^-(aq) + H_2O$$

The bicarbonate ion, one product, can neutralize additional OH^-.

Ascarite, a commercial product for removing CO_2 from an air supply, is NaOH deposited on an inert support.

$$HCO_3^-(aq) + OH^-(aq) \longrightarrow CO_3^{2-}(aq) + H_2O$$

Carbonates and Bicarbonates The important salts of carbonic acid are the water-soluble bicarbonates and carbonates of the Group IA metals and the water-insoluble carbonates of the Group IIA metals. Their reactions with acids were discussed first in Section 4.6.

Other Carbon Compounds—Cyanides, Carbides, and Carbon Disulfide

Cyanides and Hydrogen Cyanide Cyanides and covalent cyano compounds of most of the elements are known, but we will limit our study to sodium cyanide, NaCN. The cyanide ion, CN^-, is the conjugate base of a weak acid, hydrogen cyanide, HCN. Thus, any soluble cyanide will react somewhat with water to give some hydrogen cyanide and hydroxide ion (which makes the solution basic).

$[:N{\equiv}C:]^-$
cyanide ion

$$CN^-(aq) + H_2O \rightleftharpoons HCN(aq) + OH^-(aq)$$

A quantitative conversion to HCN occurs by the action of a strong acid on the cyanide ion in NaCN.

$$CN^-(s \text{ or } aq) + H^+(aq) \longrightarrow HCN(aq)$$

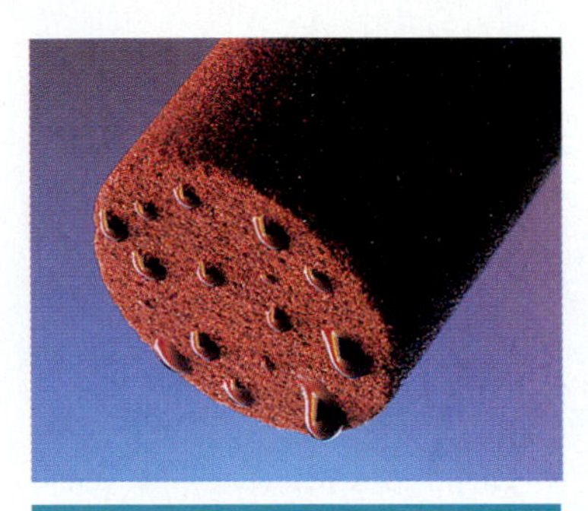

Bits of silicon carbide are incorporated into drill bits for exploratory oil drilling.

Although HCN is very soluble in water, it is easily removed as a gas by heating the solution.

HCN is an extremely poisonous gas; survivors say that it has an almond odor. Cyanide salts are equally lethal. Whatever the source of the cyanide ion, it kills by forming a complex ion with the Fe^{3+} ion, an ion essential to an enzyme system that is vital in every cell for the cell's use of oxygen. The enzyme is deactivated, however, when cyanide ions bind to its Fe^{3+} ions.

Carbides Carbides are binary compounds of carbon with metals. Some are largely saltlike or ionic, some are covalent but macromolecular, and some are called interstitial carbides.

C_2^{2-} is $[:C{\equiv}C:]^{2-}$

Among the *saltlike carbides,* some behave as if they contain the C_2^{2-} or acetylide ion. Calcium carbide is an example. In contact with water, its acetylide ion acts as a strong proton acceptor and changes to acetylene, leaving the hydroxide ion to be combined with Ca^{2+}.

$$\underset{\text{calcium carbide}}{CaC_2(s)} + 2H_2O \longrightarrow \underset{\text{acetylene}}{H{-}C{\equiv}C{-}H} + Ca(OH)_2(s)$$

Another type of saltlike carbide reacts with water to give methane, so these carbides behave as if they contained the methanide ion, C^{4-}, also a powerful proton acceptor. Magnesium carbide is an example. In contact with water, its methanide ion strips protons from water molecules, forming methane and leaving hydroxide ions to be combined with Mg^{2+} ions.

$$\underset{\text{magnesium carbide}}{Mg_2C(s)} + 4H_2O \longrightarrow \underset{\text{methane}}{CH_4(g)} + 2Mg(OH)_2(s)$$

The tungsten carbide cutting tool peels steel away from the rotating rod.

Silicon carbide, SiC, is a *covalent carbide* made by heating silicon dioxide with carbon at a high temperature (2000 to 2600 °C). The common name of the alpha form of SiC is *carborundum,* an effective abrasive for sandpapers and grinding wheels. It's almost as hard as diamond.

In *interstitial carbides,* carbon atoms occupy spaces or interstices within the lattices of metal atoms, leaving the material with many characteristics of a metal, like conductivity and luster. An industrially important example is tungsten carbide, WC, used to make high-speed cutting tools because it is exceptionally hard and chemically stable even as the tool becomes very hot during use.

Carbon Disulfide The reaction of sulfur with methane in the presence of a catalyst produces an interesting sulfur analog of carbon dioxide, namely, carbon disulfide, CS_2.

$$CH_4 + 4S \xrightarrow[Al_2O_3]{600\ ^\circ C} CS_2 + 2H_2S$$

CS_2 has a low boiling point, 46 °C, but it is also thermally unstable. Vapors of CS_2 have been known to explode on contact with a hot steam pipe or even with boiling water. Above 100 °C, CS_2 vapors spontaneously ignite in air. It is very poisonous.

SUMMARY

Occurrence of Nonmetals Just a few **nonmetal elements,** notably hydrogen and helium, contribute the greatest number of atoms to the universe.

Hydrogen Three isotopes are known—protium, deuterium, and tritium—protium (hydrogen) being the chief isotope. Deuterium gives an **isotope effect.** Hydrogen is used to make ammonia, hydrogenated vegetable oils, and chemicals. Some is used as the fuel in rockets.

Hydrides exist for nearly all elements except the noble gases. *Ionic hydrides* are hydride ion donors (and thus are reducing agents and bases). The common *covalent hydrides* tend to be proton donors or acids, but CH_4 is neither an acid nor a base; if anything, it tends to donate hydrogen *atoms.* In many covalent *metal* hydrides, the compound tends to be a hydride ion donor, but some are donors of H_2.

Oxygen Elements of the **oxygen family**—oxygen, sulfur, selenium, and tellurium—display a steady transition from a nonmetal to a metalloid. Most oxygen is obtained from liquid air. Oxygen is essential to the respiration of nearly all living things, and to the processes of decay and combustion. It is regenerated by photosynthesis, and an oxygen cycle exists in nature.

Ozone (O_3), an **allotrope** of oxygen, is a powerful oxidizing agent used by some cities to purify water.

Oxides exist for all elements except He, Ne, and Ar. When the oxidation state of the other element in the oxide is particularly high, the oxide is covalent and acidic. As the oxidation number of the other element decreases, the oxide becomes more and more ionic and basic. Between the acidic oxides of nonmetals and the basic oxides of the metals in Groups IA and IIA lie some transition metal oxides that are **amphoteric.** Their high melting points suggest that they are ionic. Only a few metal oxides are inert to acids and bases.

The **peroxides** are compounds with O—O bonds. Hydrogen peroxide, H_2O_2, is a strong oxidizing agent in both acids and bases. It is unstable, decomposing to water and oxygen.

Nitrogen The elements of the **nitrogen family** are nitrogen, phosphorus, arsenic, antimony, and bismuth. Nitrogen is the principal component of air and is obtained from it.

In compounds, nitrogen occurs in 10 oxidation states ranging from -3 to $+5$ as well as $-(1/3)$. The Haber–Bosch process converts hydrogen and nitrogen into ammonia, a fertilizer and a raw material for nitric acid and other chemicals.

Ammonia is a weak base but quantitatively neutralizes strong acids. Liquid ammonia dissolves Group IA and IIA metals to give a solvated (*ammoniated*) electron and a metal ion, but the solutions decompose eventually to metal amides (salts of the amide ion, NH_2^-) and H_2. The amide ion is a powerful base. Liquid ammonia has an acid–base chemistry analogous to that of liquid water. The ammonium ion, NH_4^+, is a proton donor and tends to form slightly acidic aqueous solutions.

Other compounds with N in a negative oxidation state include hydrazine (NH_2NH_2), hydroxylamine ($HONH_2$), and hydrazoic acid (HN_3).

Oxoacids and Oxides of Nitrogen Nitric acid is made from ammonia by the **Ostwald process.** It gives all the typical reactions of a strong, monoprotic acid, but it is more often used as an oxidizing agent.

Nitrate salts of the Group IA and IIA metals are stable beyond their melting points, which tend to be relatively low for salts, but eventually decompose to nitrites or metal oxides.

The important oxides of nitrogen are dinitrogen monoxide (N_2O, nitrous oxide), nitrogen monoxide (NO, nitric oxide), dinitrogen trioxide (N_2O_3, nitrous anhydride), nitrogen dioxide (NO_2), dinitrogen tetroxide (N_2O_4), and dinitrogen pentoxide (N_2O_5, nitric anhydride). NO_2 and N_2O_4 exist together in an equilibrium. Both NO and NO_2 are air pollutants. In aqueous base, NO_2 gives both nitrate and nitrite ions. In water alone, NO_2 reacts to give both nitric acid and nitrous acid. Nitrous acid (HNO_2) is a weak, thermally unstable acid known only in solution.

Carbon Family The members of the **carbon family**—carbon, silicon, germanium, tin, and lead—form series of hydrides, halides, and oxides with similar formulas. Carbon atoms, unlike those of the other elements in the carbon family, exist in extended chains and rings capable of holding the atoms of other nonmetals.

The chief allotropes of carbon are graphite, diamond, and members of the fullerene family. Charcoal and coke have chiefly the graphite structure. The important oxides of carbon are CO and CO_2. CO, a common fuel in industry, is also used to reduce ores to metals.

Carbon dioxide is an acidic oxide. It neutralizes strong bases and gives some carbonic acid, H_2CO_3, in water. Carbonic acid has a series of bicarbonates (of Group IA metals) and carbonates (of virtually all metals).

Carbon is present in carbon disulfide, hydrogen cyanide, the cyanide salts, and the carbides. The carbides fall roughly into three groups; saltlike carbides like CaC_2, which react with water to give acetylene; other saltlike carbides, like Mg_2C, which give methane with water; and the covalent carbides with high formula masses and resistance to any reaction with water. Examples are silicon carbide (carborundum) and interstitial metal-like carbides, such as tungsten carbide.

THINKING IT THROUGH

The goal for each of the following problems is to give you practice in thinking your way through problems. The goal is not *to find the answer itself; instead, you are only asked to assemble the available information needed to obtain the answer, state what additional data (if any) are needed, and describe how you would use the data to answer the question.*

1. If moist red litmus paper is pressed against powdered MgO, the paper turns blue. Explain.

2. The melting point of TiO_2 is 1870 °C; the melting point of OsO_4 is 40 °C. In which solid oxide are the bonds the more likely to be ionic? How do the melting points suggest the answer?

3. A white solid is either Na_2O or Na_2O_2. When a piece of red litmus paper is dipped into a freshly made aqueous solution of the white solid, its color changes from red to white. Which substance is it? How can you tell? What would happen to the red litmus paper if the solid were the other compound?

4. An element has the hypothetical symbol *E*. It has a hydride whose formula can be written either as H_2E or as EH_2. The hydride reacts as follows with water.

$$EH_2 \text{ (or } H_2E) + 2H_2O \longrightarrow E(OH)_2 + 2H_2$$

(a) Is the hydride likely to be a gas or a solid at room temperature? Explain.

(b) If the element is a representative element, in what family does it most likely belong?

5. Does the behavior of the amide ion in water tell us that this ion is a stronger or a weaker base than the hydroxide ion? Explain.

REVIEW EXERCISES[3]

Questions whose numbers are printed in color have their answers in Appendix A. Some of the more challenging questions are marked with asterisks.

Occurrence of Nonmetals

1.1 In the "big bang" theory, the initial body consisted of what particles?

1.2 What two elements are the most abundant in the universe today?

1.3 Which two elements make up most of our atmosphere and in what percentages (on a volume/volume basis)?

Hydrogen

1.4 What kinds of compounds are the chief sources of hydrogen in industry?

1.5 Give the names, atomic symbols, and mass numbers of the isotopes of hydrogen.

1.6 What is the *deuterium isotope effect?*

1.7 What are the three kinds of hydrides? Give the formula and name of an example of each kind.

[3] Several Review Exercises assume that Chapter 16 in the text ("Acid–Base Equilibria") has been studied. A few Review Exercises assume that Chapter 17 ("Solubility and Simultaneous Equilibria") and Chapter 18 ("Electrochemistry") have also been studied.

1.8 Give one physical and one chemical property in which ionic and covalent hydrides differ markedly.

1.9 Complete the following equations. If no reaction occurs, write "no reaction."

(a) $NaH(s) + H_2O \rightarrow$
(b) $CaH_2(s) + H_2O \rightarrow$
(c) $HCl(g) + H_2O \rightarrow$
(d) $Na(s) + H_2(g) \rightarrow$
(e) $Mg(s) + H_2(g) \rightarrow$

1.10 How was it determined that a compound like NaH contains the hydride ion?

1.11 Using the periodic table, which of the following hydrides are ionic? Which are molecular or covalent?

$$MgH_2,\ H_2Se,\ KH,\ HI,\ PH_3,\ CaH_2$$

1.12 What is the largest industrial use of hydrogen?

1.13 Write the overall equation for the Haber–Bosch process.

Oxygen

1.14 Give the names and atomic symbols of the members of the oxygen family.

1.15 How is the oxygen needed for industrial purposes manufactured?

1.16 Give three major *commercial* uses of oxygen.

1.17 What are the names and formulas of the allotropes of oxygen?

1.18 In what specific ways do the terms *isotope* and *allotrope* differ? (In your answer, give examples.)

1.19 How is ozone made in the lab (in general terms)?

Oxides

1.20 Consider the oxides of the representative family of elements in the periodic table. (a) Which representative family has the most ionic oxides? (b) Which family has the most basic oxides? (c) The oxides of which family are insoluble in water but can still neutralize strong acids? (d) Which family has the most acidic oxides in water?

1.21 Explain how the bonds in OsO_4 are covalent, not ionic.

1.22 Consider the action of OH^- on $[Al(H_2O)_6]^{3+}$. (a) Write the net ionic equation that describes the conversion of $[Al(H_2O)_6]^{3+}$ into $[Al(H_2O)_2(OH)_4]^-$. (b) It appears that this conversion (part a) consists of proton transfers from four of the six water molecules in the hydrated cation to OH^- ions. What makes these proton transfers possible? (Hint: They do not occur from the water molecules of a hydrated sodium ion.)

1.23 One of the reactions of molybdenum(VI) oxide is

$$MoO_3(s) + 2OH^-(aq) \longrightarrow MoO_4^{2-}(aq) + H_2O$$

From this information, are the bonds in MoO_3 likely to be covalent or ionic? How can you tell?

1.24 Sulfur trioxide is described as an *acidic oxide.* What does this mean?

1.25 $Cr(OH)_3$ is amphoteric. What does this mean? Give equations of possible reactions that illustrate this property.

Peroxides and Superoxides

1.26 What must be true, *structurally,* about any compound to call it a peroxide?

1.27 List some uses of dilute hydrogen peroxide.

1.28 Calculate the oxidation number of O in H_2O_2.

1.29 In what way is pure hydrogen peroxide dangerous?

1.30 Consider hydrogen peroxide as a weak acid. (a) Write the K_a expression for the first ionization of H_2O_2. (b) The pK_a of H_2O_2 is 11.74. What is its K_a? (c) Do a calculation to estimate the pH of a solution of 1.0 M H_2O_2.

***1.31** Hydrogen peroxide oxidizes H_2SO_3 to SO_4^{2-}. Assume that this reaction is run as a galvanic cell. (a) Write the net ionic equation for the cell reaction. (b) Based on data in this chapter and Chapter 18, calculate the standard potential of this cell. (c) How many grams of H_2O_2 are required to react with 5.25 g of Na_2SO_3 by this reaction?

1.32 Which member or members of the Group IA metals react(s) with oxygen to give chiefly (a) a peroxide? (b) An oxide? (c) A superoxide?

1.33 Describe by means of equations how potassium superoxide removes carbon dioxide from (a) dry air and (b) humid air.

Reactions Involving Oxygen and Its Compounds

***1.34** Complete and balance the following equations.

(a) $NaH(s) + O_2(g) \rightarrow$
(b) $H^-(s) + H_2O \rightarrow$
(c) $HgO(s) \xrightarrow{\text{High temperature}}$
(d) $KClO_3(s) \xrightarrow{\text{High temperature}}$
(e) $Na_2O_2(s) + H_2O \rightarrow$
(f) $Li(s) + O_2(g) \rightarrow$
(g) $H_2O_2(l) \xrightarrow{\text{High temperature}}$

1.35 Using general descriptions of chemical properties given in this chapter, figure out the products in the following reactions and write balanced molecular equations. (Some of these reactions were studied in chapters in the text.)

(a) $Na_2O(s) + H_2 \rightarrow$
(b) $Na_2O(s) + HCl(aq) \rightarrow$
(c) $Na_2O(s) + HNO_3(aq) \rightarrow$
(d) $Al_2O_3(s) + HCl(aq) \rightarrow$
(e) $Al_2O_3(s) + NaOH(aq) \rightarrow$
(f) $K_2O(s) + HCl(aq) \rightarrow$
(g) $K_2O(s) + H_2O \rightarrow$
(h) $Na_2O_2(s) + H_2O \rightarrow$
(i) $Al_2O_3(s) + HBr(aq) \rightarrow$

1.36 The addition of water to a white solid known to be either Na_2O or Na_2O_2 caused a basic solution to form. A gas evolved. When a burning match was thrust into this gas, the flame flared more brightly. What was the solid and what gas formed? Write the molecular equation.

Nitrogen

1.37 Elemental nitrogen needed for industrial purposes is obtained in what way?

1.38 With respect to enhanced oil recovery using nitrogen: (a) How does it work? (b) Why isn't (cheaper) air used instead of pure nitrogen?

1.39 Why would it be particularly unlikely for nitrogen (a) to occur in a +6 oxidation state? (b) To occur in a −4 oxidation state?

Ammonia

*1.40 The value of $\Delta H_{vaporization}$ of liquid ammonia is 1370 J g^{-1}. Suppose its value were 13.7 J g^{-1} instead. Would this make it easier or more difficult to handle liquid ammonia as a solvent? Explain.

1.41 Write net ionic equations to explain the following. (a) Ammonia in water causes an increase in pH. (b) Aqueous ammonia neutralizes hydrochloric acid. (c) Gaseous ammonia can burn to give nitrogen and water.

1.42 Write molecular equations for the reaction of $NH_3(aq)$ with each of the following acids (assumed to be in dilute solutions). This constitutes a review of the chemistry of ammonia regardless of the chapter in which the reaction was first described. (a) $HCl(aq)$, (b) $HBr(aq)$, (c) $HI(aq)$, (d) $H_2SO_4(aq)$ (as a diprotic acid), (e) $HNO_3(aq)$

1.43 When sodium dissolves in liquid ammonia, ammoniated forms of the sodium ion and the electron form. What does *ammoniated* mean?

1.44 In the liquid ammonia system of acid–base reactions, what ion is the chief proton donor and what ion is the chief proton acceptor?

1.45 How is sodium amide prepared? (Write the equation.)

1.46 What reaction will the amide ion give with the ammonium ion in a liquid ammonia solution? Write the net ionic equation.

1.47 If you add $KNH_2(s)$ to water, what will happen, chemically? (Write a molecular equation and a net ionic equation.)

1.48 At a sufficiently high temperature, $NH_4Cl(s)$ decomposes. Write the molecular equation for this decomposition.

1.49 Write the net ionic equation that explains how a solution of NH_4Cl in water has a pH less than 7.

Nitrides

1.50 When a compound is called a *nitride,* what do we know about it, structurally?

1.51 When a compound is known as a nitride of a Group IIA metal, what do we know (a) about its formula? (b) About the oxidation number of N in it? (c) About its behavior toward water?

1.52 When magnesium is ignited in air it burns with an extremely bright flame, and a white residue remains. When water is sprinkled on this residue, the odor of NH_3 can be detected. (a) How did NH_3 form? (Write the molecular equation.) (b) What other compound formed when the magnesium burned? (Write the molecular equation for its formation.)

Hydrazine

1.53 How is hydrazine prepared? (Write the molecular equation.)

1.54 What is the oxidation number of N in hydrazine?

1.55 Why is it particularly dangerous to let household bleach (of the "chlorine" type) mix with household ammonia?

1.56 When hydrazine is used as a rocket fuel, what provides the thrust? As part of the answer, write a molecular equation.

1.57 The dissolved oxygen present in any highly pressurized, high-temperature steam used in steam boilers can be extremely corrosive at the temperatures used. (The boilers are usually made of an iron alloy.) For the past several years, large steam-boiler installations have used hydrazine to remove this oxygen. (a) Write the molecular equation for the reaction by which hydrazine does this. (b) Considering the products of this reaction (part a), are they themselves harmful to the metals of the equipment? Explain.

*1.58 Hydrazine in strong acid forms both the $N_2H_5^+$ ion and the $N_2H_6^{2+}$ ion. What are the likely Lewis structures of these ions? According to VSEPR theory, what would be the likely geometry around each N atom in $N_2H_6^{2+}$?

Hydroxylamine and Hydrazoic Acid

1.59 What is the equilibrium equation that shows hydroxylamine, water, and the products of their interaction? Write the expression for K_b for hydroxylamine based on this equilibrium equation.

1.60 Using the values of K_b for ammonia and hydroxylamine given in this text, which is the stronger base?

***1.61** Based on the relative basicities of water and ammonia, which site in hydroxylamine, O or N, is likely to be the better proton acceptor? What is the Lewis structure of the protonated form of hydroxylamine?

1.62 What is the equilibrium equation that shows hydrazoic acid, water, and the products of their interaction? Write the expression for K_a for hydrazoic acid based on this equilibrium equation.

*1.63 Based on the pK_a of hydrazoic acid (4.74), will a solution of lithium azide in water test slightly acidic, slightly basic, or neutral? Explain with a net ionic equation.

1.64 What property of azides of heavy metal ions should make you careful about ever handling them?

1.65 How does valence bond theory explain the one N—N bond distance observed in the azide ion? (Draw the resonance structures.)

Nitric Acid and Nitrates

1.66 Write the molecular equations for the steps in the Ostwald process.

1.67 What is the chief commercial use of nitric acid?

1.68 After concentrated nitric acid has remained in a bottle exposed to sunlight for some time, the reagent turns from colorless to reddish brown. Write the equation for the reaction responsible for this change in color. What chemical causes the color?

1.69 Write net ionic equations for the reaction of copper with nitric acid under the following conditions. (a) Dilute nitric acid is used and is reduced to $NO(g)$. (b) Concentrated nitric acid is used and is reduced to $NO_2(g)$.

1.70 From the ways in which sodium nitrate and potassium nitrate respond to heat above 500 °C, what can we learn about the relative thermal stabilities of sodium nitrite and potassium nitrite?

1.71 What characterizes the nitrates of Group IIA metals with respect to their thermal stability?

1.72 The decomposition of ammonium nitrate takes different directions according to the temperature at which it is heated. Write equations for these modes at (a) 200–260 °C and (b) over 300 °C. (c) What do these facts *alone* possibly tell us about the thermal stability of N_2O and the identities of the products of the thermal decomposition of N_2O?

Oxides of Nitrogen and Nitrous Acid

1.73 One reason why aerosol cans carry a "Do Not Incinerate" warning is found in the pressure–temperature law of gases. When the propellant is N_2O, there is an additional reason. What is it?

1.74 Explain why N_2O_4 is diamagnetic but NO_2 is paramagnetic.

1.75 How does nitrous acid decompose in water? (Write the equation.)

1.76 The pK_a of HNO_2 (at 18 °C) is 3.35. (a) Calculate K_a. (b) Write the equilibrium equation on which K_a is based.

1.77 What does the addition of $NaNO_2$ to water do to the pH—raise it, lower it, or leave it unchanged? Write a net ionic equation that explains your answer.

1.78 How is dinitrogen pentoxide prepared? (Write the equation.)

1.79 Draw the Lewis structure of HNO_3.

1.80 Deduce the Lewis structure of the nitronium ion.

1.81 Give the name and formula of the oxide of nitrogen that does the following. (a) Gives nitric acid when dissolved in water. (b) Forms N_2O_4 when cooled. (c) Can be made by heating ammonium nitrate. (d) Is unstable in air, being oxidized to NO_2. (e) Forms in an automobile cylinder by a reaction between nitrogen and oxygen. (f) Is unstable and readily breaks up into two other oxides of nitrogen. (g) Is a reddish-brown, poisonous gas. (h) Gives the same reactions as a mixture of NO and NO_2. (i) Is used by some auto racers to get more power out of the combustion of the fuel. (j) Are paramagnetic. (k) Is a solid at room temperature. (l) Is recycled in the Ostwald process. (m) Forms from the decomposition of nitrous acid in water. (n) Reacts with aqueous sodium hydroxide to give a mixture of sodium nitrite and sodium nitrate. (o) Is responsible for the characteristic color of heavy smog.

Carbon and Its Inorganic Compounds

1.82 Give the names and the atomic symbols of the elements in the carbon family.

1.83 Describe the similarities and differences among coke, charcoal, and graphite. What is activated charcoal used for, and how does it work? What is a fullerene? Why is graphite slippery?

1.84 Write equations for the industrial production of CO that gives water gas.

1.85 How can small amounts of CO be made in the lab? (Write an equation.)

1.86 How is CO used in the metallurgical industry? (Give an equation of one example.)

1.87 How does CO work as a poison?

1.88 Write the equation for the preparation of lime.

1.89 In general terms, how is portland cement made and what substances are in it?

1.90 What happens chemically when portland cement sets (in general terms)?

1.91 Dry ice sublimes, so how is it possible to make *liquid* CO_2?

1.92 Although relatively little of the carbon dioxide in an aqueous solution of this gas is present as carbonic acid, the solution can be treated as a neutralizer of base as if all of the carbon dioxide were in this form. Explain, using equations.

1.93 All metal carbonates and bicarbonates react alike with strong acids. What are the net ionic equations for these reactions? Write one equation for carbonates and one for bicarbonates (assuming that both are in solution).

1.94 How is carbon disulfide made? (Write an equation.)

1.95 A solution of potassium cyanide in water has a pH above 7. Write a net ionic equation that explains this.

1.96 Using a net ionic equation, describe what happens when hydrochloric acid is poured into a concentrated solution of sodium cyanide.

1.97 How does the cyanide ion act as a poison (in general terms)?

1.98 We discussed three types of carbides. Give specific examples, including both names and formulas. What is carborundum?

1.99 Write the equation for the reaction (if any) of each compound with water.
(a) WC (b) CaC_2 (c) SiC (d) Mg_2C

Formulas, Names, and Reactions of Substances

1.100 Write the formula of each. (Remember that the formula of an ion must include the kind and amount of charge.) (a) hydride ion, (b) potassium amide, (c) ammonium sulfate, (d) ozone, (e) amide ion, (f) sodium oxide, (g) sodium peroxide, (h) sodium hydride, (i) ammonium nitrate, (j) lithium amide, (k) sodium nitride, (l) nitric acid, (m) sodium amide, (n) nitrous acid, (o) hydrogen peroxide, (p) hydrazine, (q) nitric oxide, (r) sodium nitrite, (s) nitrous oxide, (t) nitric anhydride, (u) hydroxylamine, (v) hydrazoic acid, (w) calcium carbide, (x) magnesium carbide

***1.101** Complete and balance the following equations. In many parts, you are expected to figure out the formulas of products from general statements about chemical properties made in the chapter. Write molecular equations.

(a) $Zn(s) + HCl(aq) \rightarrow$
(b) $Ca(s) + H_2(g) \rightarrow$
(c) $NaH(s) + H_2O \rightarrow$
(d) $Mg(s) + HCl(aq) \rightarrow$
(e) $KH(s) + H_2O \rightarrow$
(f) $HgO(s) \xrightarrow{\text{Heat}}$
(g) $Na_2O(s) + H_2O \rightarrow$
(h) $MgO(s) + HBr(aq) \rightarrow$
(i) $KClO_3(s) \xrightarrow{\text{Heat}}$
(j) $Al_2O_3(s) + HBr(aq) \rightarrow$
(k) $Al_2O_3(s) + KOH(aq) \rightarrow$
(l) $CrO_3(s) + H_2O \rightarrow$
(m) $H_2O_2(aq) \xrightarrow{\text{Heat}}$
(n) $K_2O_2(s) + H_2O \rightarrow$
(o) $CO_2(aq) + H_2O \rightleftharpoons$
(p) $CO_2(aq) + OH^-(aq) \rightarrow$
(q) $CaC_2(s) + H_2O \rightarrow$
(r) $MgC_2(s) + H_2O \rightarrow$
(s) $CaCO_3(s) \xrightarrow{\text{Heat strongly}}$
(t) $NaCN(aq) + HCl(aq) \rightarrow$

***1.102** Complete and balance the following equations. In many parts, you are expected to figure out the formulas of products from general statements about chemical properties made in the chapter. Write molecular equations.

(a) $NH_3(aq) + HCl(aq) \rightarrow$
(b) $NH_4Br(s) \xrightarrow{\text{Heat}}$
(c) $NH_4Cl(aq) + NaOH(aq) \rightarrow$
(d) $NH_4Cl(am) + KNH_2(am) \xrightarrow{NH_3(l)}$
(e) $(NH_4)_2Cr_2O_7(s) \xrightarrow{\text{Heat}}$
(f) $Li_3N(s) + H_2O \rightarrow$
(g) $NH_3(aq) + NaOCl(aq) \rightarrow$
(h) $N_2H_4(l) + O_2(g) \rightarrow$
(i) $NH_2OH(aq) + HCl(aq) \rightarrow$
(j) $HN_3(aq) + LiOH(aq) \rightarrow$
(k) $MgCO_3(s) \xrightarrow{\text{Heat strongly}}$
(l) $HBr(aq) + KCN(aq) \rightarrow$
(m) $HCO_3^-(aq) + OH^-(aq) \rightarrow$

***1.103** Complete and balance the following equations. In many parts, you are expected to figure out the formulas of products from general statements about chemical properties made in the chapter. Write molecular equations.

(a) $NO(g) + O_2(g) \xrightarrow{\text{Heat}}$
(b) $KOH(aq) + HNO_3(aq) \rightarrow$
(c) $N_2O(g) \xrightarrow{\text{Heat}}$
(d) $NO_2(g) + H_2O \rightarrow$
(e) $N_2O_3(g) \xrightarrow{\text{Heat}}$
(f) $N_2O_5(s) + H_2O \rightarrow$
(g) $N_2O_4(g) \xrightarrow{\text{Heat}}$
(h) $HNO_3(l) + P_4O_{10}(s) \rightarrow$
(i) $NaNO_3(s) \xrightarrow{>500\ ^\circ C}$
(j) $CO_2(aq) + KOH(aq) \rightarrow$
(k) $N_2O_3(g) + NaOH(aq) \rightarrow$
(l) $NaNO_2(aq) + HCl(aq) \xrightarrow{0\ ^\circ C}$
(m) $NaNO_2(aq) + HCl(aq) \xrightarrow{\text{Heat}}$

Additional Problems

***1.104** Consider the formation and properties of hydrazine. (a) Write the thermochemical equation for the formation of one mole of liquid hydrazine from its elements in their standard states. The value of ΔH_f° for $N_2H_4(l)$ is $+50.6$ kJ mol^{-1}. (b) Toward its own elements, is hydrazine *thermodynamically* stable or unstable? How can you tell? (c) Pure hydrazine is described as being *kinetically* stable with respect to its decomposition to its elements. What does this mean? (d) The value of ΔG_f° for $N_2H_4(l)$ is $+149.2$ kJ mol^{-1}. Calculate ΔS_f° for $N_2H_4(l)$, and comment on the meaning of its algebraic sign in relationship to the thermochemical equation for its formation.

***1.105** In a basic solution, hydrazine can be changed to N_2 and H_2O, with $E^\circ = -1.16$ V (25 °C). (a) Write the half-reaction *in the accepted manner* for this change. (b) Using electrochemical data, could hydrazine reduce $Ag^+(aq)$ to $Ag(s)$? Write the cell equation and calculate the cell potential.

***1.106** The fact that concentrated nitric acid, acting on copper, is changed to NO_2, not NO (which forms when *dilute* nitric acid is used), suggests that excess concentrated nitric acid can *oxidize* NO to NO_2. (a) What would be the half-reaction for the reaction of HNO_3 with NO to give NO_2? (b) What would be the resulting cell reaction? (c) What would be the resulting value of E_{cell}°? Does this value suggest that the reaction of HNO_3 with NO would be spontaneous? Explain. (d) What would the Nernst equation suggest as to the effect of a high concentration of HNO_3 on giving a favorable value to E?

***1.107** The bond order of the N—N bond in N_2O is about 2.5 instead of the value of 3 in N_2. How does molecular orbital theory explain this lower bond order?

***1.108** Write the molecular and net ionic equations for the reactions of hydrochloric acid with each of the following compounds. If no reaction occurs, write "no reaction."

(a) $NaOH(aq)$
(b) $NaHCO_3(aq)$
(c) $Na_2CO_3(aq)$
(d) $KOH(aq)$
(e) $K_2CO_3(aq)$
(f) $KHCO_3(aq)$
(g) $CaCO_3(s)$
(h) $Ca(OH)_2(s)$
(i) $Mg(OH)_2(s)$
(j) $MgCO_3(s)$
(k) $Na_2S(aq)$
(l) $K_2SO_3(aq)$
(m) $NaCl(aq)$
(n) $KBr(aq)$
(o) $LiCN(aq)$
(p) $Pb(NO_3)_2(aq)$
(q) $AgNO_3(aq)$
(r) $Ca(s)$
(s) $NaNO_2(aq)$
(t) $K_2SO_4(aq)$
(u) $Mg(s)$
(v) $NaC_2H_3O_2(aq)$
(w) $NaNH_2(s)$
(x) $KN_3(s)$
(y) $NH_3(aq)$
(z) $CO_2(aq)$

***1.109** Write molecular and net ionic equations for the reaction of hydrobromic acid with each of the substances listed in Review Exercise 1.108.

***1.110** Write molecular and net ionic equations for the reaction of nitric acid with each of the substances listed in Review Exercise 1.108.

***1.111** Write molecular and net ionic equations for the reaction of sulfuric acid with each of the substances listed in parts (a) through (f) and parts (k) through (z) of Review Exercise 1.108. Assume that only sulfate compounds, not hydrogen sulfate compounds, form.

1.112 Suppose that the volume of nitrogen at STP released by the operation of an auto airbag is 20.0 L. (a) How many grams of sodium azide decomposed? (b) How many grams of sodium form? (c) How many grams of iron(III) oxide are needed to react with the sodium that is produced?

Sulfur, an element studied in this chapter, sometimes forms near the rims of volcanic vents, such as this crater of the Sierra Negra, Ecuador.

Chapter 2

Simple Molecules and Ions of Nonmetals: Part II

2.1 SULFUR

Sources

Sulfur, called "brimstone" by the ancients, occurs in underground deposits as a brittle, bright yellow, nonmetallic element (Figure 2.1). It is also a component of sulfide and sulfate minerals and is present in various forms in natural gas, petroleum, and coal. Underground sulfur is brought to the surface by the Frasch process, which provides about a third of the annual United States sulfur production (Figure 2.2). In the Frasch process, superheated steam is pumped into the deposit, causing the sulfur to melt and gather into pools. Then hot air and steam pressure forces molten sulfur to the surface.

In the mid 1990s, annual sulfur production in the United States was on the order of 11–12 million tons.

Over half of the sulfur produced in the U.S. is isolated from its compounds. For example, sulfur is obtained by using sulfur dioxide to oxidize the hydrogen sulfide present in natural gas and petroleum, an operation that also removes a poisonous and smelly substance.

$$2H_2S(g) + SO_2(g) \longrightarrow 3S(s) + 2H_2O$$

The needed sulfur dioxide is made by the controlled oxidation of some of the hydrogen sulfide in the oil or gas.

$$2H_2S(g) + 3O_2(g) \longrightarrow 2SO_2(g) + 2H_2O$$

FIGURE 2.1
Orthorhombic sulfur, S_α.

Isotopes of Sulfur Of the four isotopes of sulfur (see the margin table), sulfur-32 predominates, but the proportions of isotopes vary more for sulfur at different locations than for most common elements. The sulfur in the sulfate ions in ocean water, for example, is enriched slightly in sulfur-34, while the sulfur in terrestrial sulfide ores is slightly depleted in this isotope. (Bacteria in terrestrial sediments that consume SO_4^{2-} and produce H_2S use $^{32}SO_4^{2-}$ slightly more rapidly than $^{34}SO_4^{2-}$. Runoff water that reaches the ocean, therefore, contains a slightly disproportionate excess of $^{34}SO_4^{2-}$ than $^{32}SO_4^{2-}$.) The differences are not great, but they prevent us from giving an atomic mass for sulfur any more precise than 32.066.

Isotope	% Abundance
^{32}S	95.02
^{33}S	0.75
^{34}S	4.21
^{35}S	0.02

The Principal Allotropes of Sulfur

Be sure to distinguish between *isotope* and *allotrope*.

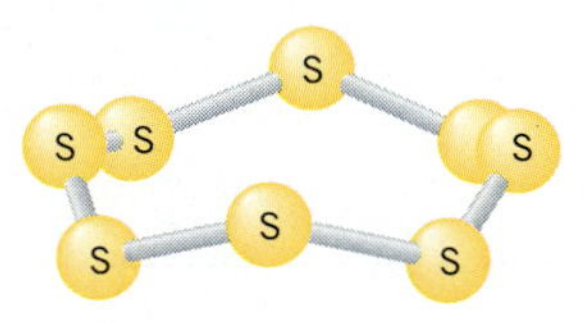

S_8 crown configuration

We'll study four of the sulfur allotropes—there are several—to illustrate how molecular size and shape influence crystalline appearances. The two most common allotropes are different crystalline forms of *cyclooctasulfur,* S_8, namely, orthorhombic and monoclinic sulfur. The S_8 molecules in both exist as crown-like rings of eight S atoms. *Orthorhombic sulfur,* or S_α, is thermodynamically the most stable of all sulfur allotropes, but it behaves most interestingly when heated. Only if *rapidly* heated does S_α have a sharp melting point, 112.8°C. If kept unmelted at a temperature of 95.5 °C, the crown S_8 molecules of S_α slowly take up new positions in the crystals. The crystal structure changes to that of *monoclinic sulfur,* S_β, which occurs in long needles (Figure 2.3). Like S_α, S_β has a sharp melting point (119 °C) only if *rapidly* heated. Slow heating allows time for some of its S_8 ring molecules to open up and so give a mixture that can start to melt at a lower temperature. (Recall that mixtures have depressed melting points; see Section 12.7.)[1] However, if left several weeks at ordinary temperature, S_β changes back to the orthorhombic form, S_α.

A third allotrope, *plastic sulfur,* forms when molten sulfur is rapidly cooled. The broken rings in molten sulfur form new S—S bonds without much order or regularity. New chains and rings of varying size form. Neat, regularly shaped crystals cannot be built from them, and the product is amorphous (Figure 2.4). On standing, however, plastic sulfur also reverts slowly to S_α, the thermodynamically most stable allotrope.

When sulfur is heated beyond its melting point, some interesting changes occur that further illustrate the effect of molecular size on physical properties (Figure 2.5). Freshly melted sulfur, in which the molecules are still mostly S_8 rings, exists first as a straw-yellow, transparent, and mobile fluid. At 159 °C, however, a sharp transition occurs, and the density, surface tension, viscosity, and other physical properties of the liquid dramatically change. Between 160 °C and 195 °C, the viscosity of molten sulfur increases by 100,000-fold! What happens is that increasing numbers of S_8 rings open to give chains *whose ends join,* and the chains may become as many as 200,000 sulfur atoms long. Such huge molecules intertangle and do not easily slip by each other as the liquid flows. The fluid, therefore, becomes extremely viscous between 160 °C and 190 °C and stays viscous until the temperature is well above 200 °C. Now, enough thermal energy is available to break the chains into much smaller lengths, and the molten sulfur becomes less and less viscous until the liquid—by now dark red—boils at 444.60 °C.

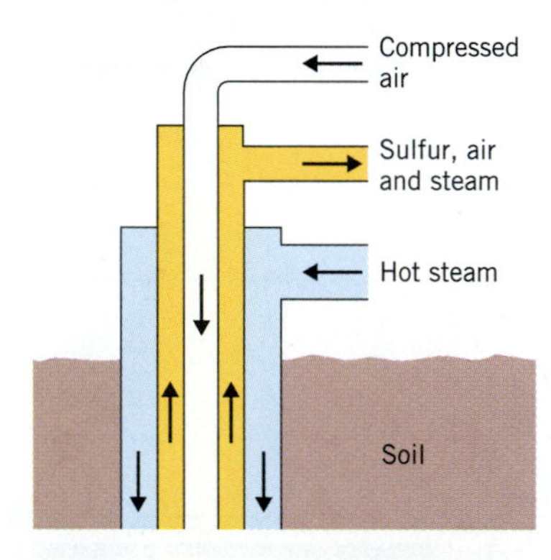

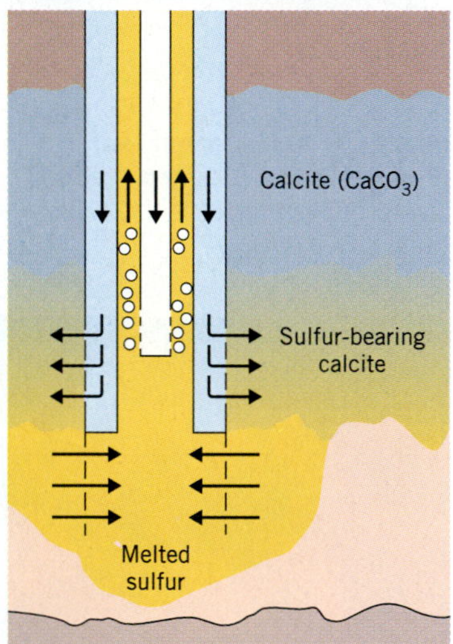

FIGURE 2.2

The Frasch process for extracting sulfur from deep deposits.

Chemical Properties of Sulfur—An Overview When heated so as to cause the breaking of S—S bonds in S_8 rings, sulfur can be made to combine *directly* with nearly all of the elements. The exceptions are the noble gases and six others—gold, platinum, iridium, tellurium, nitrogen, and iodine.

Bonds from S to these six elements can be formed indirectly.

Sulfur has a high ionization energy (999.30 kJ mol^{-1}) and a relatively high electron affinity (+200 kJ mol^{-1}), which both tell us that sulfur does not readily give up electrons. Consequently, sulfur can exist in *positive* oxidation states largely when combined with elements of similarly high electron affinities, like fluorine, oxygen, and chlorine. Otherwise, sulfur tends to be in a

[1] Unless noted otherwise, cross-references to Sections, Special Topics, and *Chemicals in Use* refer to the accompanying text, *Chemistry: The Study of Matter and Its Changes,* 2nd edition (1996), by Brady and Holum, not to this supplement. Some material from text chapters 15 to 19 is assumed for this supplement.

TABLE 2.1 The Oxidation States of Sulfur

Oxidation State	Examples
−2	H_2S, HS^-, S^{2-}, CH_3SH
−1	S_2^{2-}
0	S
+4	SO_2, HSO_3^-, SO_3^{2-}
+6	SO_3, H_2SO_4, HSO_4^-, SO_4^{2-}

FIGURE 2.3
Monoclinic sulfur, S_β.

reduced or negative oxidation state, as in hydrogen sulfide and metal sulfide salts. Table 2.1 summarizes the various oxidation states that sulfur ordinarily has, together with some examples of typical compounds.

The +4 Oxidation State of Sulfur

At an ignition temperature of 250 °C to 260 °C, sulfur burns with a bright blue flame to give gaseous sulfur dioxide.

For simplicity, we will revert to writing sulfur as S, not as S_8.

$$S(s) + O_2(g) \longrightarrow SO_2(g)$$

The sharp odor of a freshly lit match is caused by SO_2.

In both environmental and economic terms, this reaction is probably the most important reaction of sulfur.

Sulfur dioxide (BP, −10 °C; MP, −72 °C) has a sharp, acrid, and choking odor. It is quite soluble in water, forming a 12% solution at 15 °C.

The chief use of sulfur dioxide is to make sulfuric acid, but enormous quantities of SO_2 never make it this far. They enter and pollute the atmosphere, making SO_2 a major contributor to "acid rain" (see *Chemicals in Use* 10).

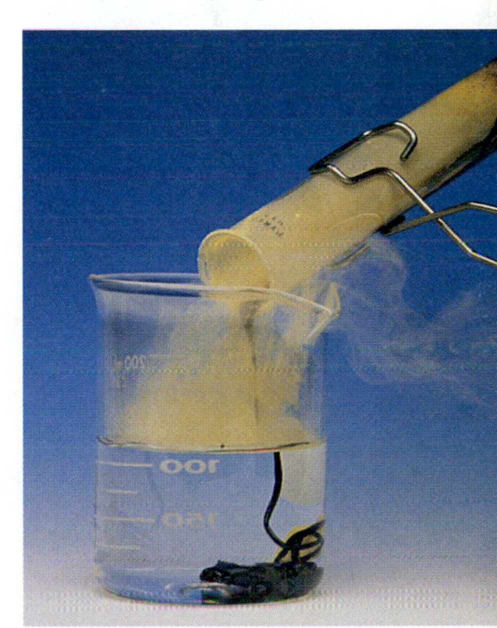

FIGURE 2.4
Plastic sulfur forms when molten sulfur is poured into water.

Sulfurous Acid Sulfur dioxide dissolves in water by forming hydrates, $SO_2 \cdot nH_2O$, where n varies with concentration, temperature, and pH. The hydrates are in equilibrium with some hydronium ion and hydrogen sulfite ion, HSO_3^-, whose presence has long been explained simply in terms of $H_2SO_3(aq)$, sulfurous acid. Actual molecules of this species—H_2SO_3—have never been detected in or out of water. Nevertheless, for convenience in writing chemical equations, the formula H_2SO_3 is widely used for the solute in aqueous sulfur dioxide. Traditionally, the first ionization of sulfurous acid is therefore represented as follows.

$$H_2SO_3(aq) \rightleftharpoons H^+(aq) + HSO_3^-(aq)$$

Actually, however, it should be represented as

$$SO_2 \cdot nH_2O(aq) \rightleftharpoons H_3O^+(aq) + HSO_3^-(aq)$$

The second ionization of sulfurous acid is the ionization of the hydrogen sulfite ion, HSO_3^-, which does exist in solution.

$$HSO_3^-(aq) \rightleftharpoons H^+(aq) + SO_3^{2-}(aq)$$

For sulfurous acid at 25 °C,
$K_{a_1} = 1.2 \times 10^{-2}$
$K_{a_2} = 6.6 \times 10^{-8}$

Hydrogen Sulfites, Disulfites, and Sulfites Chemical supply houses sell a product called *sodium bisulfite* (or sodium hydrogen sulfite), $NaHSO_3$, but the material in the bottle is mostly sodium disulfite, $Na_2S_2O_5$ (sometimes called sodium metabisulfite). In an acidified solution, however, $NaHSO_3$ and

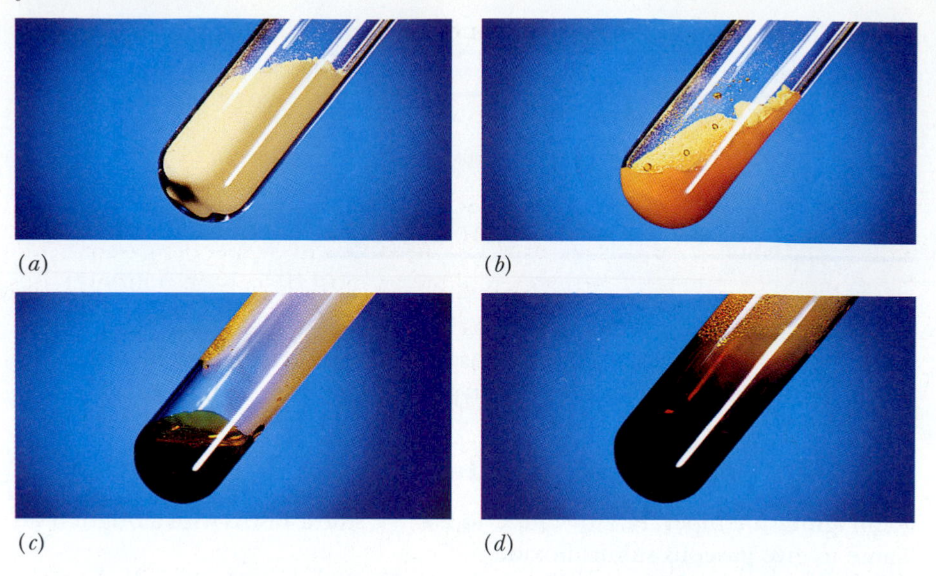

FIGURE 2.5

Changes in sulfur as it is heated. (*a*) Crystalline sulfur, S_α. (*b*) Molten sulfur just above its melting point. (*c*) Molten sulfur, just below 200 °C, is dark and viscous. (*d*) Boiling sulfur is dark red and no longer viscous.

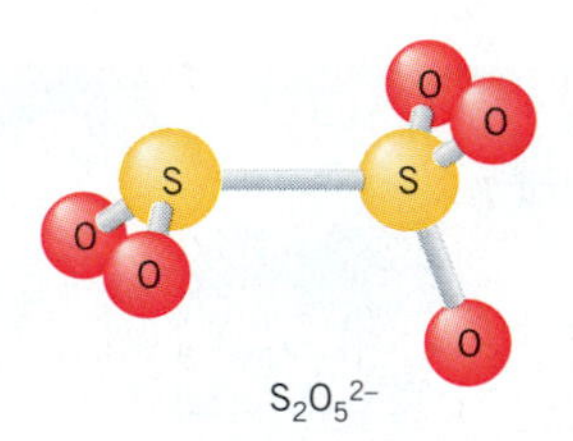

$Na_2S_2O_5$ are chemically equivalent because of the following equilibrium.

$$2HSO_3^-(aq) \rightleftharpoons \underset{\text{disulfite ion}}{S_2O_5^{2-}(aq)} + H_2O$$

A *solution* of sodium hydrogen sulfite can be made directly by passing SO_2 into aqueous Na_2CO_3.

$$2SO_2(g) + Na_2CO_3(aq) + H_2O \longrightarrow \underset{\substack{\text{sodium}\\\text{hydrogen sulfite}\\\text{(sodium bisulfite)}}}{2NaHSO_3(aq)} + CO_2(g)$$

However, complex events occur if we try to isolate solid $NaHSO_3$ from this solution by boiling off the water, a common strategy for the isolation of solutes. Some Na_2SO_3, some SO_2, and some $Na_2S_2O_5$ form, but no solid $NaHSO_3$.

Sodium sulfite is made by the reaction of sodium carbonate with aqueous sodium hydrogen sulfite followed by the removal of the water.

$$2NaHSO_3(aq) + Na_2CO_3(aq) \longrightarrow \underset{\text{sodium sulfite}}{2Na_2SO_3(aq)} + CO_2(g) + H_2O$$

The reaction is really just an acid–base neutralization. The resulting solution can be evaporated to dryness to give sodium sulfite, either as the heptahydrate ($Na_2SO_3 \cdot 7H_2O$) or as the anhydrous form, depending on the temperature.

Sodium sulfite is the most important sulfite salt. Hundreds of thousands of tons are made per year throughout the world, chiefly for use in the manufacture of paper pulp. A small amount is used as a food additive to prevent bacterial decomposition. Some people are extremely allergic to sulfites, however, and in the early 1990s the U.S. government required that all products with sulfites disclose their use and carry a warning on the label.

At low pH values, any of the following equilibria involving SO_2 or the sulfites shifts to the right, so the particular species present in solution depends on the pH.

$$SO_3^{2-}(aq) + H^+(aq) \rightleftharpoons HSO_3^-(aq)$$

$$HSO_3^-(aq) + H^+(aq) \rightleftharpoons H_2SO_3(aq)$$

The sulfites are mild reducing agents. In acid, sulfurous acid participates with the sulfate ion in the following half-reaction.

$$SO_4^{2-}(aq) + 4H^+(aq) + 2e^- \rightleftharpoons H_2SO_3(aq) + H_2O \qquad E° = +0.172 \text{ V}$$

The low value of $E°$ means that the reverse reaction, where H_2SO_3 gives up electrons and functions as a reducing agent, can occur with many compounds —anything that is a stronger oxidizing agent than SO_4^{2-}. However, a relatively concentrated aqueous solution of SO_2 is unstable with respect to gaseous sulfur dioxide. Hence, the following equilibrium must be kept in mind when a sulfite is used as a reducing agent in an acidic solution.

$$H_2SO_3(aq) \rightleftharpoons H_2O + SO_2(g)$$

The escape of $SO_2(g)$ means a loss of the reducing ability from the solution (and also means that a fume hood must be used).

People with severe sulfite allergies should read the ingredient lists on food and beverage containers.

The +6 Oxidation State of Sulfur

The chief compounds having sulfur in its +6 state are sulfur trioxide, sulfuric acid, and its salts. Sulfur trioxide is made by the catalytic oxidation of sulfur dioxide by oxygen. (Sulfur dioxide itself does not burn in air; a catalyst is needed.)

Sulfur Trioxide Sulfur trioxide (BP, 44.6 °C; MP, 16.86 °C) exists in the gaseous and liquid states as molecules of SO_3 in equilibrium with molecules of a trimer, S_3O_9. The physical properties of sulfur trioxide, therefore, are for the equilibrium mixture, which is colorless. Solid sulfur trioxide, which exists entirely as the trimer when perfectly dry, easily sublimes. For simplicity, we will represent sulfur trioxide as SO_3.

Sulfur trioxide gas has an even sharper, more choking odor than sulfur dioxide, and the conversion of SO_3 to sulfuric acid in any moist tissue makes SO_3 particularly dangerous. This property of SO_3 also contributes to the overall problem of acid rain.

The oxidation of sulfur dioxide to sulfur trioxide takes place to a *slight* extent in any sunlit atmosphere polluted with SO_2 and dust particles.

$$2SO_2 + O_2(g) \longrightarrow 2SO_3(g) \qquad \Delta H° = -191.2 \text{ kJ}$$

Although the reaction is exothermic, it occurs very slowly in the absence of a catalyst. Lucky us! If the oxidation of SO_2 occurred instantaneously in air, like the oxidation of NO to NO_2, either the use of sulfur-bearing coal or oil would be banned outright, or elaborate and very costly steps would be needed to remove *all* SO_2 from smokestack gases.

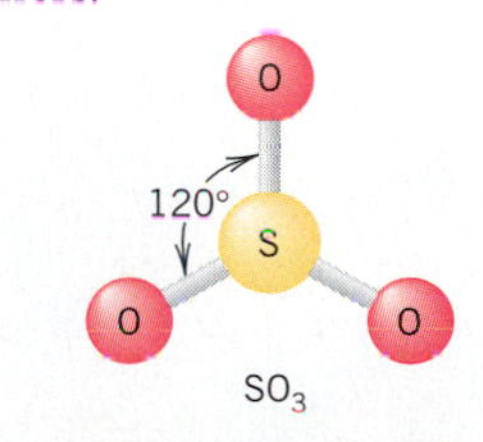

Sulfur trioxide is an acidic oxide, and it reacts violently and very exothermically with water to form sulfuric acid.

$$SO_3(g) + H_2O \longrightarrow H_2SO_4(aq) \qquad \Delta H° = -880 \text{ kJ}$$

Sulfur trioxide also neutralizes aqueous solutions of Groups IA and IIA hydroxides, forming salts of the hydrogen sulfate ion, HSO_4^-, or of the sulfate ion, SO_4^{2-}, depending on the mole proportions taken. For example,

$$SO_3(g) + NaOH(aq) \longrightarrow NaHSO_4(aq)$$

or

$$SO_3(g) + 2NaOH(aq) \longrightarrow Na_2SO_4(aq) + H_2O$$

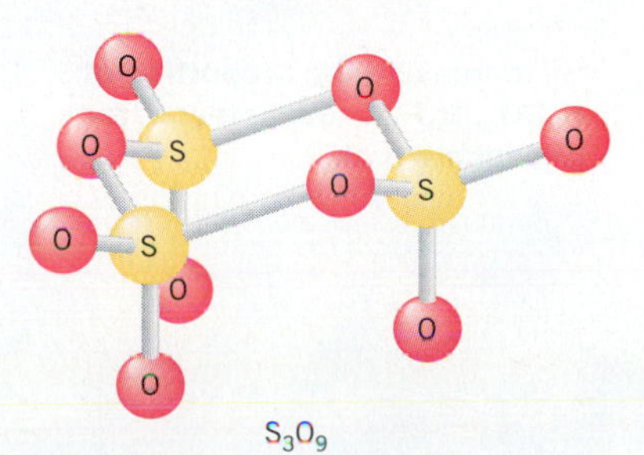

Sulfur trioxide also neutralizes bicarbonates and carbonates. For example,

$$SO_3(g) + Na_2CO_3(aq) \longrightarrow Na_2SO_4(aq) + CO_2(g)$$

In terms of *moles,* ammonia production at 1 × 10^{12} mol/year outdistances sulfuric acid production at 4 × 10^{11} mol/yr.

Sulfuric Acid Sulfuric acid is manufactured by the **contact process,** the name denoting *contact* between SO_2 and a catalyst during one step. All steps are exothermic. In the first, sulfur is burned to give sulfur dioxide.

$$S(s) + O_2(g) \longrightarrow SO_2(g) \qquad \Delta H° = -297 \text{ kJ}$$

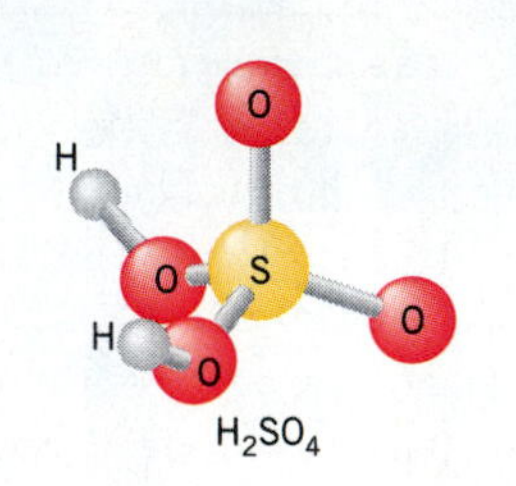

The sulfur dioxide is passed, together with oxygen, over heated beds of the catalyst, vanadium pentoxide, and the SO_2 is oxidized to SO_3.

$$2SO_2(g) + O_2(g) \xrightarrow{V_2O_5} 2SO_3(g) \qquad \Delta H° = -191.2 \text{ kJ}$$

The SO_3 is now bubbled into concentrated sulfuric acid, which reacts with and traps the gas as disulfuric acid, $H_2S_2O_7$ (sometimes called pyrosulfuric acid).

$$SO_3(g) + H_2SO_4(l) \longrightarrow H_2S_2O_7(l)$$

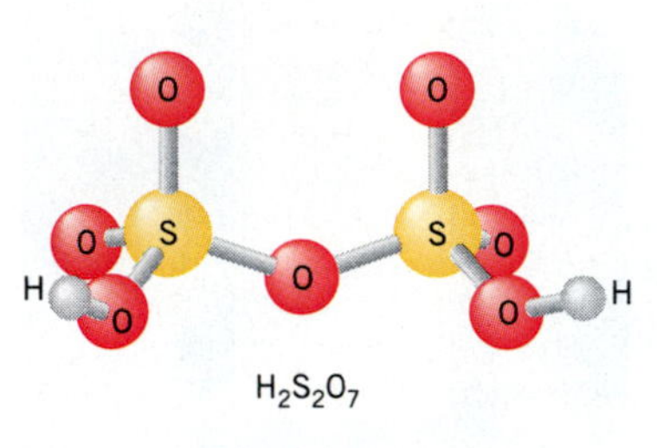

Water is now added, and disulfuric acid breaks down to sulfuric acid.

$$H_2S_2O_7(l) + H_2O \longrightarrow 2H_2SO_4(l)$$

The overall value of the enthalpy change for the last two steps is $\Delta H° = -130$ kJ. (The reason the direct addition of SO_3 to liquid water is not used is that the SO_3 would contact water vapor first and produce a fine mist of sulfuric acid; this would be hard to trap without its escaping to the atmosphere.)

96% H_2SO_4 is 18 *M* H_2SO_4.

The addition of water to $H_2S_2O_7$ can be controlled to give a final product with a concentration of 95 to 98% H_2SO_4, the usual form of commercially available concentrated sulfuric acid. [The disadvantage of 100% H_2SO_4 is that it melts at 10.4 °C (51 °F), and is therefore susceptible to solidification at cool temperatures and so is harder to handle.]

Sulfuric acid, a strong diprotic acid, gives all of the usual reactions of strong acids with active metals, and their oxides, hydroxides, bicarbonates, and carbonates. Whether one or both protons from H_2SO_4 are used is largely a matter of the chemist's choice of mole proportions. Concentrated sulfuric acid is a dangerous chemical when incautiously used (see Special Topic 2.1).

Chemical Uses of Sulfuric Acid Sulfuric acid has several advantages over other acids for industrial uses. It is the least expensive of the strong acids. It is the only strong acid that can be prepared and shipped in an almost pure form. At or near room temperature it is stable and nonvolatile. It is not a vigorous oxidizing agent, unless heated. It is a potent dehydrating agent, being even powerful enough to gather the pieces of water molecules out of sugar molecules ($C_{12}H_{22}O_{11}$) leaving only a spongy mass of carbon (Figure 2.6).

About 70% of the annual sulfuric acid production is consumed to make fertilizers, mostly in a process that converts insoluble phosphate rock into more soluble phosphate forms. Phosphate rock in many mines is principally $Ca_5(PO_4)_3F$, which is very insoluble in water and not of much use as a source of the phosphate ion needed by growing plants. However, phosphate rock reacts with sulfuric acid as follows to give a mixture called *superphosphate,* which is much more soluble in water.

By increasing the proportion of H_2SO_4, H_3PO_4 can be made from phosphate rock.

$$2Ca_5(PO_4)_3F + 7H_2SO_4 + 17H_2O \longrightarrow \underbrace{7CaSO_4\cdot 2H_2O + 3Ca(H_2PO_4)_2\cdot H_2O}_{\text{"superphosphate"}} + 2HF$$

SPECIAL TOPIC 2.1 / Working with Concentrated Sulfuric Acid

In the early 1990s, sulfuric acid was being manufactured in the United States at an annual rate of about 44 million tons (400×10^9 mol), making it the country's largest volume chemical (in terms of tonnage). Sulfuric acid is used in so many ways to make commercial products that the per capita use of this substance is a good index for the industrial activity of a nation and its standard of living.

Sulfuric acid is thick and syrupy, so concentrated sulfuric acid tends to cling to surfaces. If you spill a drop on your skin, expect a chemical burn and a blister regardless of how rapidly you flush the site with cold water. Start flushing the area with water even if you are not sure if the drop is concentrated acid. The only experience most people have with sulfuric acid is with automobile lead storage batteries. Be sure to handle or dispose of such batteries very carefully.

All of the reactions of concentrated sulfuric acid are very exothermic, including simply diluting it with water. Quickly removable protective gloves as well as safety glasses must be worn when dispensing concentrated sulfuric acid. Give yourself space so that no one will inadvertently bump you as you pour this acid. Never pick up a jug without providing some support at its base, because jug handles have been known to break off. Never carry a jug without setting it inside a special rubber bucket large enough to contain the acid fully if the jug should break (see Figure 1*a*). And always remember the first rule about diluting concentrated solutions—always add the concentrated solution slowly, with stirring, to the water. Never add water to the concentrated solution. If you do the latter, so much heat can be generated when the first water hits the concentrated acid that some material will almost surely boil and spatter out of the container.

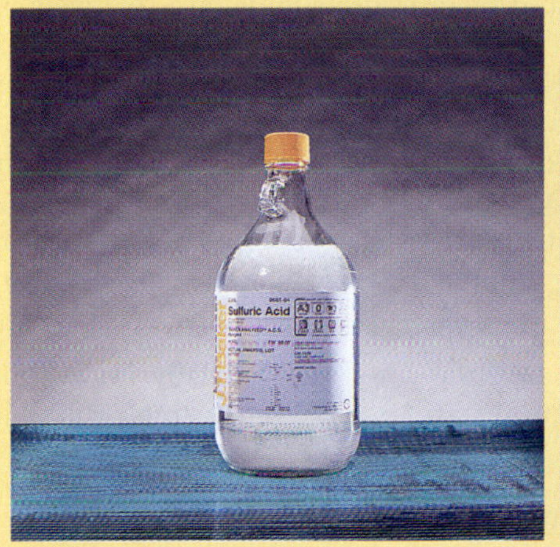

(*a*)

(*b*)

(*a*) Concentrated sulfuric acid. (*b*) Bucket for carrying concentrated sulfuric acid, which will hold the entire contents if the bottle breaks.

FIGURE 2.6

The effect of concentrated sulfuric acid on sugar. The same quantity of sugar, $C_{12}H_{22}O_{11}$, had been placed in both beakers. Concentrated sulfuric acid was then added to the beaker on the right. The acid was able to draw the components of water from the sugar molecules, whose formula could be rewritten as $C_{12}(H_2O)_{11}$, and thus leave behind a spongy mass of elemental carbon (with sulfuric acid sticking to it).

The gypsum ($CaSO_4 \cdot 2H_2O$) is left in the product. The hydrogen fluoride is trapped because, otherwise, it would be a serious air pollutant. At air levels as low as 0.1 ppm HF is toxic to certain plants, causing chlorosis and edge and tip burn. In growing animals, HF interferes with bone development, causing bone spurs and oversized bones.

***Chlorosis* is the loss of chlorophyll, the green pigment in plants necessary for photosynthesis.**

In a 1.0 *M* solution of acetic acid ($K_a = 1.8 \times 10^{-5}$), a typical *weak* acid, the percentage dissociation is less than 1%. The HSO_4^- ion, a much stronger acid than acetic acid ($K_a = 1.0 \times 10^{-2}$), is over 30% ionized in 0.10 *M* $NaHSO_4$.

Hydrogen Sulfates and Sulfates In an aqueous solution, the hydrogen sulfate ion is a moderately strong Brønsted acid existing in an equilibrium with the sulfate ion.

$$HSO_4^-(aq) \rightleftharpoons H^+(aq) + SO_4^{2-}(aq) \qquad K_a = 1.0 \times 10^{-2}\ (25\ ^\circ C)$$

Thus, when a *solid* acid is needed, $NaHSO_4$ is often used. By and large, the common hydrogen sulfate salts are those of the Group IA metals.

The Group IA sulfates (Table 2.2) are all ionic, high-melting solids and all are quite soluble in water. Their aqueous solutions generally have pH values of around 7. Chemists use anhydrous sodium sulfate as an inert drying agent because it combines strongly with water to form hydrates.

The Group IIA sulfates (Table 2.2) either decompose or melt at high temperatures and, except for beryllium and magnesium sulfate, are relatively insoluble in water. Most are available as hydrates. Gypsum, $CaSO_4 \cdot 2H_2O$, is present in alabaster and is used in plaster board. Another hydrate, plaster of Paris, $(CaSO_4)_2 \cdot H_2O$, can be used to make plaster casts.

The strong ability of $MgSO_4$ to form hydrates makes this an important drying agent, sold under the name Drierite. The thermal stability of magnesium sulfate (below 400 °C) makes it possible to regenerate spent Drierite granules indefinitely simply by baking them to drive off the water of hydration.

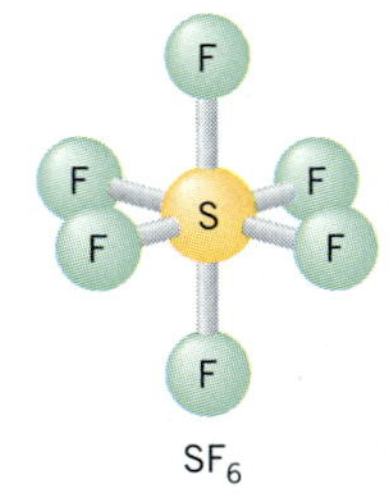

SF_6

Sulfur Hexafluoride, SF_6 Sulfur hexafluoride, an especially interesting compound of sulfur in the +6 oxidation state, is made by burning sulfur in fluorine.

$$S(s) + 3F_2(g) \longrightarrow SF_6(g)$$

Sulfur hexafluoride, alone among the many fluorides of sulfur, is exceptionally stable. At room temperature it is a colorless, odorless, tasteless, nontoxic, nonflammable, water-insoluble gas. It is so stable and it resists electrical currents so well that it is used as an insulating gas in electrical generators and

TABLE 2.2 Hydrogen Sulfates and Sulfates of Group IA and IIA Metals

Group IA Salts					
Hydrogen Sulfates		Sulfates		Group IIA Sulfates[a]	
Formula	Melting Point (°C)	Formula	Melting Point (°C)	Formula	Melting Point (°C)
$LiHSO_4$	170	Li_2SO_4	845	$BeSO_4$	Dec. 550–600
$NaHSO_4$	315	Na_2SO_4	885	$MgSO_4$	Dec. 1124
$KHSO_4$	214	K_2SO_4	1067	$CaSO_4$	1450
$RbHSO_4$	—	Rb_2SO_4	1060	$SrSO_4$	1605
$CsHSO_4$	Dec.[b]	Cs_2SO_4	1019	$BaSO_4$	Dec. >1600

[a] The hydrogen sulfates do not exist as free solids.

[b] Decomposes without melting.

switches. Few metals attack SF_6, but boiling sodium metal breaks it down to NaF and Na_2S.

Other Oxoacids of Sulfur and Their Salts At least 10 oxoacids or oxoacid salts of sulfur are known, but four of the acids are unstable and not known as pure compounds. Only their salts have been prepared. We will study only one type, the **thiosulfates.**

The *thiosulfate ion,* $S_2O_3^{2-}$, which forms many stable salts, can be thought of as the anion of the unstable thiosulfuric acid, $H_2S_2O_3$. Sodium thiosulfate is made by the reaction of sulfur with hot, aqueous sodium sulfite.

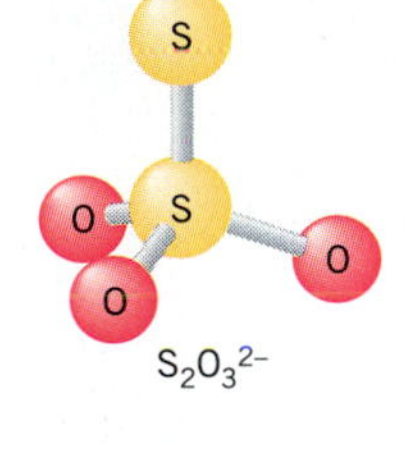

$$S(s) + Na_2SO_3(aq) \xrightarrow{\text{heat}} \underset{\text{sodium thiosulfate}}{Na_2S_2O_3(aq)}$$

If you acidify the resulting solution to try to make thiosulfuric acid, $H_2S_2O_3$, a complex mixture of products forms instead.

The thiosulfate ion is a moderate reducing agent, as the standard reduction potential indicates for its conversion to the tetrathionate ion, $S_4O_6^{2-}$.

$$\underset{\text{tetrathionate ion}}{S_4O_6^{2-}(aq)} + 2e^- \rightleftharpoons 2S_2O_3^{2-}(aq) \qquad E° = +0.169 \text{ V}$$

For example, the thiosulfate ion reduces iodine to the iodide ion as follows.

$$2S_2O_3^{2-}(aq) + I_2(aq) \longrightarrow S_4O_6^{2-}(aq) + 2I^-(aq)$$

Chlorine, a stronger oxidizing agent than iodine, oxidizes the thiosulfate ion further, namely, to the hydrogen sulfate ion.

$$S_2O_3^{2-}(aq) + 4Cl_2(g) + 5H_2O \longrightarrow 2HSO_4^-(aq) + 8H^+(aq) + 8Cl^-(aq)$$

Tap water from chlorinated water supplies kills fish, so aquarium owners use this reaction to remove chlorine.

The bleaching industry has used sodium thiosulfate in this reaction to destroy excess chlorine bleaches in fibers.

The pentahydrate of sodium thiosulfate—$Na_2S_2O_3 \cdot 5H_2O$—is the familiar *hypo* that photographers use as a fixing agent when they develop film (see Special Topic 3.3, next chapter).

The −2 Oxidation State of Sulfur

Sulfur occurs in the −2 state in several sulfide ores as well as in hydrogen sulfide and its salts.

Some sulfide ores:

FeS_2	iron pyrite
Cu_2S	chalcocite
ZnS	sphalerite
HgS	cinnabar
PbS	galena
$CuFeS_2$	chalcopyrite

Hydrogen Sulfide and Related Compounds Hydrogen sulfide is a gas whose odor of rotten eggs nearly everyone knows. It is a more dangerous poison than hydrogen cyanide, being lethal at a concentration of only 100 ppm and very rapid acting. Fortunately, its distinctive odor is detected at a concentration of only 0.02 ppm, so people quickly move to cleaner air. Because H_2S desensitizes the nose's sense of smell, however, it is unwise to judge its concentration by odor.

Hydrogen sulfide can be made by the action of a strong acid on any number of metal sulfides, like iron(II) sulfide.

$$FeS(s) + 2HCl(aq) \longrightarrow H_2S(g) + FeCl_2(aq)$$

A geologist in the field knows that when a drop of hydrochloric or sulfuric acid is placed on a mineral suspected of being a sulfide and the odor of rotten eggs develops, the mineral is a sulfide. However, if no odor develops, the opposite conclusion cannot be drawn, because some sulfide minerals, like CuS and HgS, are acid-insoluble sulfides.

Hydrogen sulfide (BP, − 50 °C; MP, − 83 °C) forms a saturated solution in water at 25 °C with a concentration of about 0.1 *M*. Its chief use in the laboratory is to detect and identify those metal ions that form sulfide precipitates having characteristic colors and solubilities. For this use, the most common preparation of aqueous hydrogen sulfide is by the reaction of water with thioacetamide.

$$\underset{\text{thioacetamide}}{CH_3\overset{\overset{\displaystyle S}{\|}}{C}NH_2} + H_2O \longrightarrow \underset{\text{acetamide}}{CH_3\overset{\overset{\displaystyle O}{\|}}{C}NH_2} + H_2S$$

For $H_2S(aq)$ at 20 °C, $K_{a_1} = 1.3 \times 10^{-7}$ and $K_{a_2} = 1.3 \times 10^{-14}$.

Hydrogen sulfide is a weak, diprotic acid that has two kinds of salts, hydrogen sulfides like NaHS and sulfides like Na_2S. Their distinctive properties are those of their anions, HS^- and S^{2-}. Both anions are Brønsted bases, the sulfide ion being strong enough a proton acceptor that when a soluble sulfide like Na_2S dissolves in water, its S^{2-} ion reacts quantitatively with water to give HS^-.

$$S^{2-}(s) + H_2O \longrightarrow HS^-(aq) + OH^-(aq)$$

The addition of a strong acid either to a hydrogen sulfide or to a sulfide salt liberates hydrogen sulfide. For example,

$$HCl(aq) + NaHS(aq) \longrightarrow H_2S(g) + NaCl(aq)$$

$$2HCl(aq) + Na_2S(aq) \longrightarrow H_2S(g) + 2NaCl(aq)$$

Because of these reactions, all experiments that might put hydrogen sulfide into the air must be done at an efficient fume hood.

Other Binary Sulfur–Halogen Compounds Sulfur forms numerous binary compounds with the halogens—six compounds with fluorine alone. We have studied SF_6 earlier.

Of the many binary sulfur–chlorine compounds, two are commercially important—disulfur dichloride (S_2Cl_2) and sulfur dichloride (SCl_2). Disulfur dichloride, made by the reaction of chlorine with molten sulfur, is a foul-smelling, toxic liquid (BP, 138 °C) used in the conversion of latex rubber into a more wear-resistant form. Sulfur dichloride, SCl_2, a red liquid (BP, 59 °C), is made from S_2Cl_2 by the further action of chlorine in the presence of a catalyst (e.g., $FeCl_3$). Both sulfur chlorides react with water to give complex mixtures of products. Binary compounds of sulfur with bromine and iodine are also known.

Other Members of the Sulfur Family—Selenium, Tellurium, and Polonium

Polonium has more isotopes (27) than any other element and all are unstable, radioactive substances.

In going down the oxygen family in Group VIA, we move from two nonmetals—oxygen and sulfur—to two metalloids (semimetals)—selenium and tellurium—and end with a radioactive metal—polonium. Yet there are many similarities among these elements, and we will look very briefly at some. All below sulfur are relatively rare.

Selenium, like sulfur, has many allotropes and some involve crownlike rings

TABLE 2.3 Acid Dissociation Constants of Some Group VIA Oxoacids

Name	Formula	K_{a_1} (25 °C)
Sulfurous acid	H_2SO_3	1.2×10^{-2}
Selenous acid	H_2SeO_3	4.5×10^{-3}
Tellurous acid	H_2TeO_3	3.3×10^{-3}
Sulfuric acid	H_2SO_4	Large
Selenic acid	H_2SeO_4	Large
Telluric acid	H_6TeO_6	2×10^{-8}

of eight selenium atoms, Se_8. Commercial selenium, however, is more an amorphous, vitreous solid (and black in color). Tellurium has only one crystalline form. Metallic polonium exists in a simple cubic structure, the only element known to do so.

All members of the oxygen family form hydrides of the general formula H_2Z, where Z represents any member of the family. H_2Se and H_2Te are gases at room temperature and known to be weak acids in water (although stronger than H_2S). H_2Po is a volatile liquid.

Selenium and tellurium form salts with the Group IA metals—M_2Se and M_2Te (where M is the metal ion). They also form salts with the Group IIA metals—MSe and MTe. These salts are, of course, similar in formula to the sulfide salts.

Selenium and tellurium, like sulfur, form dihalides, dioxides, and trioxides. The oxides are chemically similar to those of sulfur, reacting with water, for example, to give acids of similar formulas and acid strengths (Table 2.3). Both selenous and tellurous acids are white solids. Selenic acid resembles sulfuric acid in acidity, but telluric acid differs in formula and is a weak acid.

2.2 PHOSPHORUS

Phosphorus was discovered in 1669 by a Hamburg alchemist, Hennig Brandt, who distilled the residue from boiled-down, well-putrified urine, condensed the vapors under water, and found he had something that glowed in the dark and burst into flame in warm air. All from urine! It must certainly have made Brandt's reputation and given hope to those seeking to make gold from baser things. The properties that Brandt observed gave the element its name, after the Greek roots *phos,* "light," and *phoros,* "bringing."

Sources and Allotropes

Phosphorus occurs entirely as one isotope, phosphorus-31, but in several allotropic forms. It is present in nature only in its compounds. They occur in the animal, plant, and mineral kingdoms, almost always as one or another variation of the phosphates. Mixed calcium phosphates, the *apatites,* are the most common mineral family of phosphorus. They have the general formula $[Ca_3(PO_4)_2]_3 \cdot CaX_2$, where X is F in fluorapatite, Cl in chloroapatite, and OH in hydroxyapatite. Organic phosphates are very widespread in living systems. The chemicals of heredity, DNA and RNA, are examples.

Allotropes of Phosphorus Phosphorus has several allotropes, but nitrogen has none, a difference related to atomic size. A triple bond—one sigma and two pi bonds—can connect two nitrogen atoms, but phosphorus atoms normally form only P—P single bonds. Double bonds between phosphorus atoms are uncommon. Let us see why.

Nitrogen atoms have much smaller radii than phosphorus atoms (70 pm *versus* 120 pm). Two nitrogen atoms can therefore approach each other more closely during bond formation. This allows for effective *sideways* overlap of the valence shell $2p$ orbitals, which is necessary for pi bonds (see Section 8.6). As a result, nitrogen forms strong pi bonds and so is able to exist solely as diatomic molecules with N≡N triple bonds. Because the phosphorus atom is much larger and must use p orbitals at level 3 instead of level 2, these would overlap poorly, side to side, to give a pi bond. Phosphorus, therefore, does not form strong pi bonds. Each phosphorus, instead, normally uses three *single* bonds (sigma bonds), one to each of three neighboring phosphorus atoms. Because different geometries can accommodate such sigma bonds, phosphorus has several allotropes.

Phosphorus sometimes uses its $3d$ orbitals in the formation of bonds, for example, to make the sp^3d hybrid orbitals in PCl_5.

What Hennig Brandt so picturesquely stumbled onto was *white phosphorus*, P_4, one of the less stable, yet most common allotropes of phosphorus (Figure 2.7). It is a waxy, white solid that melts at 44 °C, boils at 280 °C, and dissolves in nonpolar organic solvents like carbon disulfide and benzene. It is *extremely* toxic, and any kind of direct body contact with it must be avoided.

In air, white phosphorus spontaneously bursts into flame at about 35 °C (Figure 2.8), so it has been used by military forces in incendiary devices. In moist air and below 35 °C, the slower reaction of white phosphorus with oxygen causes the glowing that Brandt observed. This phenomenon is now called *chemiluminescence* but an earlier name, *phosphorescence*, more closely tied the observed glowing to the element itself. Why does white phosphorus have these unusual properties?

The four atoms in the P_4 molecule are at the corners of a regular tetrahedron, and the bond angles are only 60°, much smaller than normal for a trivalent atom. The small angles mean poor overlap between atomic orbitals, so the P—P bonds are weak in P_4. This probably explains why white phosphorus is so easily attacked by atmospheric oxygen and should not be stored except under water and out of contact with air.

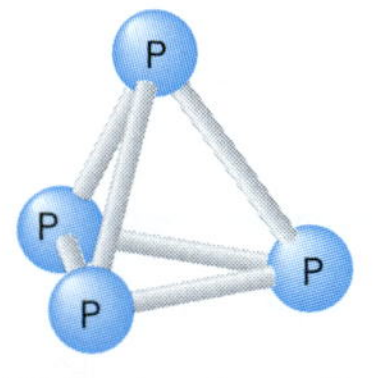

White phosphorus, P_4

White phosphorus is produced today on the scale of a few million tons per year by the reduction of calcium phosphate by carbon, with silicon dioxide being used to capture the calcium as a (molten) calcium silicate slag. The

FIGURE 2.7

The two main allotropes of phosphorus. White phosphorus is on the left and red amorphous phosphorus is on the right.

reduction requires a high temperature even though it is exothermic. For the formation of P_4 by the following reaction, $\Delta H° = -3060$ kJ.

$$2Ca_3(PO_4)_2(s) + 10C(s) + 6SiO_2(s) \xrightarrow{1400-1500\ °C} P_4(g) + 10CO(g) + 6CaSiO_3(l)$$

The combustion of the CO made here helps to provide the energy for the reaction.

Most manufactured phosphorus is used to make phosphoric acid.

FIGURE 2.8

When pure oxygen is bubbled into contact with white phosphorus (P_4), stored under water, the phosphorus bursts into flame.

Red Phosphorus White phosphorus changes spontaneously to another common allotrope, *red amorphous phosphorus,* when it is carefully heated in the absence of air between 270 and 300 °C. Its structure is complex, as the term *amorphous* implies.

Red amorphous phosphorus (MP, 600 °C) is essentially nontoxic, can be stored in air, and is altogether far safer than white phosphorus. Both red and white phosphorus are involved in the chemistry of matches (see Special Topic 2.2). For simplicity in further discussion, we will use the monatomic symbol, P, for phosphorus.

Phosphine

Binary compounds of phosphorus with all elements except bismuth, antimony, and those of Group 0 (noble gases) are known. *Phosphine,* PH_3, the most stable binary hydride of phosphorus, is an extremely poisonous gas with a garlic odor. It can be made by the hydrolysis of calcium phosphide, which is produced when calcium and phosphorus are heated.

$$\underset{\text{calcium phosphide}}{Ca_3P_2(s)} + 6H_2O \longrightarrow \underset{\text{phosphine}}{2PH_3(g)} + 3Ca(OH)_2(s)$$

Although PH_3 is the structural analog of NH_3, there are few chemical similarities. The P—H bond is nonpolar, because the electronegativities of P and H are the same, ruling out hydrogen bonds from PH_3 to water molecules. Although the hydrogen bonds from NH_3 molecules to those of water are weak, they nevertheless help ammonia to be over 2500 times more soluble in water than phosphine. Unlike ammonia, phosphine does not readily react either as a proton acceptor or a proton donor. When ignited, phosphine burns in air to give orthophosphoric acid. (Sometimes impurities make phosphine burn spontaneously.)

$$PH_3(g) + 2O_2(g) \longrightarrow \underset{\text{orthophosphoric acid}}{H_3PO_4(l)}$$

The chief use of phosphine is in the manufacture of other substances that make cotton cloth flame-resistant.

Halogen Compounds of Phosphorus

Twelve phosphorus halides are known, occurring in three sets according to the general formulas PX_3, P_2X_4, and PX_5 where X is F, Cl, Br, or I. Many mixed halides with different halogen atoms present are also known.

We will look briefly only at two chlorides, PCl_3 and PCl_5, then generalize about others, and also learn about phosphorus oxychloride, $POCl_3$. We will be interested mainly in two kinds of reactions, those with water and with oxygen, because these are common reactants (and are both present in humid air).

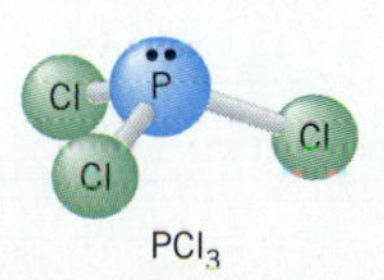

PCl_3

SPECIAL TOPIC 2.2 / Matches

We take matches so much for granted that it's hard to imagine what life was like before safe and inexpensive matches were developed.

The "safety match" was invented in 1855 by J. E. Lundstrom (Sweden). Glued to the match head is a mixture mostly of potassium chlorate—a strong oxidizing agent—and antimony(III) sulfide. Glued to the striking surface is a mixture of red phosphorus, antimony(III) sulfide, a little iron(III) oxide, and powdered glass. Striking the match produces some heat that converts a tiny trace of red phosphorus to white phosphorus, which instantly ignites. The heat ignites the chemicals in the match head, and their short blaze ignites the wood or paper of the matchstick.

In the "strike-anywhere match," invented in 1889 by H. Sévene and E. Cahan (France), the match head contains potassium chlorate, tetraphosphorus trisulfide (P_4S_3), ground glass, and the oxides of zinc and iron, all bound by glue. The scratch gives enough heat to initiate a violent (but small-scale) reaction between $KClO_3$ and P_4S_3, and its heat ignites the matchstick.

PCl_3 and PCl_5 are made by the direct chlorination of phosphorus.

$$2P(s) + 3Cl_2(g) \longrightarrow 2PCl_3(l)$$

$$2P(s) + 5Cl_2(g) \longrightarrow (PCl_4)^+(PCl_6)^-(s)$$

Phosphorus trichloride, or phosphorus(III) chloride, is a colorless liquid (BP, 76 °C) that exists in all phases as a covalent substance. It is used to make organic compounds either of phosphorus or of chlorine.

Phosphorus pentachloride, or phosphorus(V) chloride, is a whitish solid (MP, 167 °C) with a structure at the ionic–covalent borderline (Figure 2.9). In the solid state and in concentrated solutions in polar solvents, like nitromethane (CH_3NO_2), PCl_5 exists as a pair of ions—$(PCl_4)^+(PCl_6)^-$. However, in the liquid and gaseous states as well as in nonpolar solvents such as carbon tetrachloride or benzene, phosphorus(V) chloride occurs as molecules of PCl_5. We will simplify our further study by using PCl_5 as the formula.

Phosphorus pentachloride, a fairly large-scale industrial chemical (20 to 30 thousand tons per year, worldwide), is a raw material for making a number of inorganic and organic phosphorus compounds. Some are macromolecular and have unusual resistance to flames and solvents and remain flexible at low temperatures, properties that are useful for making fuel hoses, seals, and gaskets for vehicles operating at very low temperatures. (The principal barrier to these applications is the cost of the chemicals involved.)

Both the trichloride and the pentachloride of phosphorus react with water, even the moisture in humid air, to give oxoacids and hydrogen chloride. In humid air, hydrogen chloride forms whitish mists of hydrochloric acid droplets, and so the phosphorus halides are said to *fume* in moist air. They must be stored in well-capped bottles.

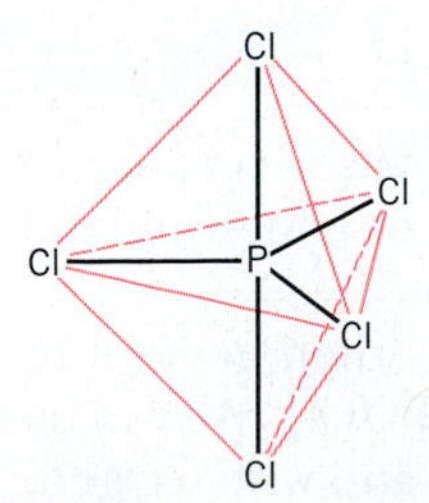

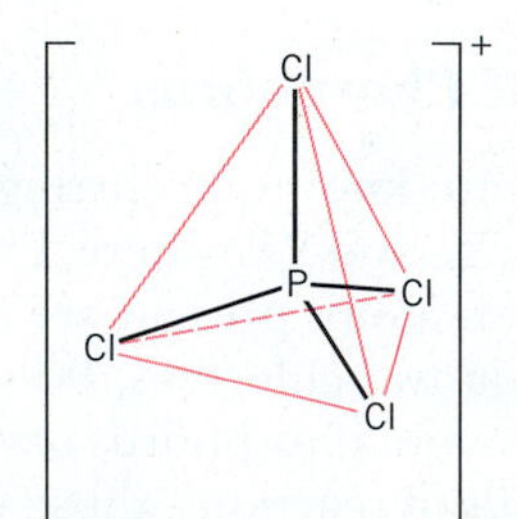

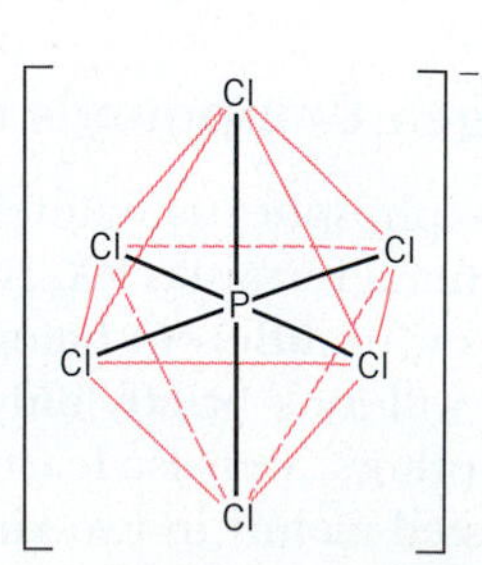

FIGURE 2.9

Phosphorus(V) chloride has the structure on the left in the gaseous state but exists as a pair of ions (right), in the solid state.

$$2PCl_3(l) + 6H_2O \longrightarrow \underset{\text{phosphorous acid}}{2H_3PO_3(aq)} + 6HCl(aq)$$

$$PCl_5(s) + 4H_2O \longrightarrow \underset{\text{phosphoric acid}}{H_3PO_4(aq)} + 5HCl(aq)$$

This reaction is violent.

Exactly analogous reactions are shown by the tri- and pentabromides.

At room temperature, the phosphorus(III) halides are easily oxidized by pure oxygen to oxohalides, POX_3. The rate of the reaction is slower in air.

$$2PCl_3(l) + O_2(g) \longrightarrow \underset{\text{phosphorus oxychloride}}{2POCl_3(l)}$$

Because phosphorus is in its highest oxidation state (+5) in its pentahalides, the pentahalides cannot be oxidized in a similar way.

Phosphorus oxychloride, like the phosphorus chlorides, fumes in moist air, and reacts very readily with water to give phosphoric acid and hydrochloric acid.

$$POCl_3(l) + 3H_2O \longrightarrow H_3PO_4(aq) + 3HCl(aq)$$

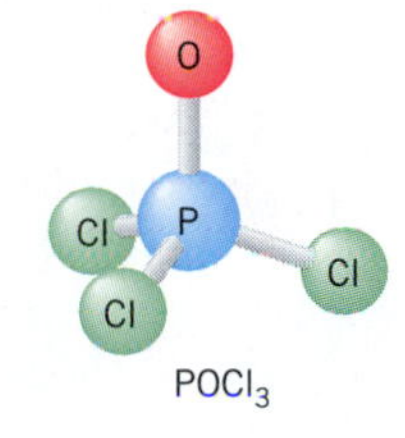

Oxides and Oxoacids of Phosphorus

Between 80 and 90% of all elemental phosphorus produced is used to manufacture phosphoric acid, the first step being combustion. People who have seen fireworks displays have witnessed how spectacularly phosphorus burns in air (Figure 2.10). The product is a cloud of incandescent, white, powdery *tetraphosphorus decaoxide,* P_4O_{10}.

$$4P(s) + 5O_2(g) \longrightarrow P_4O_{10}(s)$$

FIGURE 2.10

Fireworks displays show the combustion of phosphorus in beautiful action. Seen here is a display over the Statue of Liberty, New York harbor.

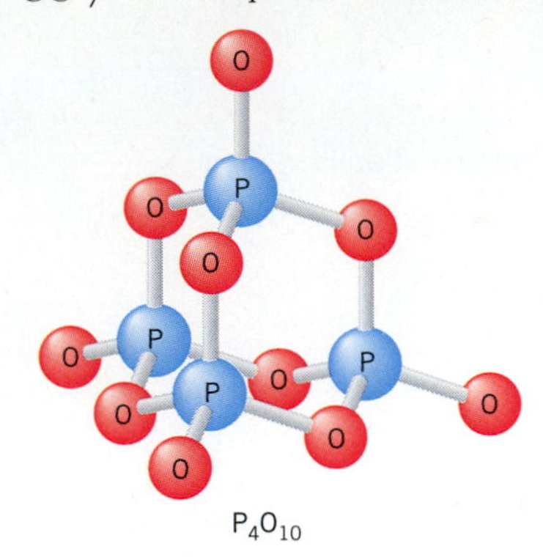

P_4O_{10}

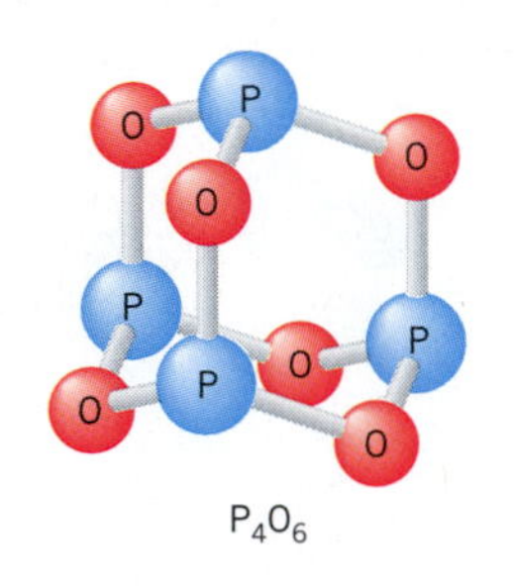

P_4O_6

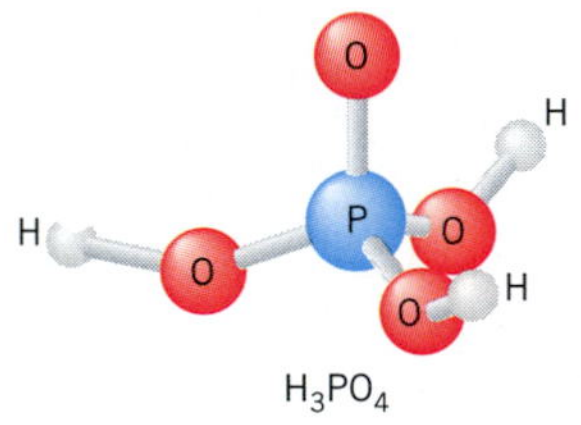

H_3PO_4

85.0% H_3PO_4 is 14.7 M H_3PO_4.

For H_3PO_4,

$K_{a_1} = 7.1 \times 10^{-3}$
$K_{a_2} = 6.3 \times 10^{-8}$
$K_{a_3} = 4.5 \times 10^{-13}$

In insufficient air, the oxidation goes to the + 3 oxidation state of phosphorus in *tetraphosphorus hexaoxide,* P_4O_6.

$$4P(s) + 3O_2(g) \longrightarrow P_4O_6(s)$$

The most notable chemical property of P_4O_{10} is its extremely exothermic reaction with water. When it is dusted onto water, the system hisses and spits as the reaction occurs. P_4O_{10} is one of the best desiccating agents (water-removing agents) available to chemists, and it must be stored in well-capped bottles. Its reaction with water is also its major industrial use. The problem with this reaction—ours, not nature's—is that we cannot write a simple equation. Any one or a mixture of a family of three phosphoric acids can form depending on the temperature and the degree to which water has been taken up.

HPO_3	$H_4P_2O_7$	H_3PO_4
metaphosphoric acid (empirical formula)	diphosphoric acid (pyrophosphoric acid)	orthophosphoric acid (phosphoric acid is the usual name)

By studying these formulas, you can see that orthophosphoric acid is simply metaphosphoric acid to which H_2O has been added, and diphosphoric acid is two orthophosphoric acids minus a water molecule. The structural formulas of the oxoacids of phosphorus are complicated by the presence of extended chains or rings, which we will soon study. Each acid has its own family of salts.

Phosphoric Acid Orthophosphoric acid (MP, 42 °C) is usually called simply *phosphoric acid,* because it is the common type of phosphoric acid in commerce. (We will use this name, too.) Phosphoric acid is sold not as a solid but as an 85% solution in water. Extensive hydrogen bonding makes both molten phosphoric acid and the 85% aqueous solution viscous, syrupy liquids.

Phosphoric acid is a moderately strong, triprotic acid. One of its minor uses is to give a tart taste to manufactured soft drinks, but its largest use is to make "triple superphosphate" fertilizer. The reaction is just like the reaction of sulfuric acid with phosphate rock (page 48)—conversion of the phosphate salt to the more soluble dihydrogen phosphate salt—but no calcium sulfate forms to dilute the fertilizer's essential potency as a supplier of "phosphate." Hence the name "triple superphosphate."

$$\underset{\text{phosphate rock}}{Ca_5(PO_4)_3F} + 7H_3PO_4 + H_2O \longrightarrow \underset{\text{"triple superphosphate"}}{5Ca(H_2PO_4)_2 \cdot H_2O} + HF$$

Soil without phosphorus (in the form of ions of phosphoric acid) is barren. Plant growth absolutely requires phosphate, because so many key compounds in cells—plants and animals alike—are organophosphates.

The wheat on the left benefits from phosphate fertilizer, which has been denied the wheat on the right in this test plot.

Salts of Phosphoric Acid Three kinds of salts of H_3PO_4 are known—those of the *dihydrogen phosphate ion,* $H_2PO_4^-$, the *monohydrogen phosphate ion,* HPO_4^{2-}, and the *phosphate ion,* PO_4^{3-}. The sodium, potassium, calcium, and ammonium salts are the most common, and they are commercially available either as anhydrous salts or as a large variety of different kinds of hydrates. When these salts are used, we have to be aware of their acid–base properties, because the hydrolysis of their anions can affect the pH of a solution (see Section 16.5). The three anions of phosphoric acid are all Brønsted bases, but they differ widely in base strength, as their base ionization constants show.

$$PO_4^{3-} > HPO_4^{2-} > H_2PO_4^-$$

$$K_b \quad 2.2 \times 10^{-2} \quad 1.6 \times 10^{-7} \quad 1.4 \times 10^{-12}$$

Remember, the *larger* the K_b, the stronger is the base.

An aqueous solution of sodium phosphate contains the strongest base of the three anions, namely, PO_4^{3-}. The solution has a pH above 7 because OH^- is generated by the following equilibrium involving PO_4^{3-} and its conjugate acid, HPO_4^{2-}.

$$PO_4^{3-}(aq) + H_2O \rightleftharpoons HPO_4^{2-}(aq) + OH^-(aq)$$

In a 0.01 *M* solution of Na_3PO_4, for example, the pH is 11.6, which is roughly equivalent to the basicity of 0.004 *M* NaOH. (You would definitely not want to expose your skin or eyes to solutions of Na_3PO_4.)

Salts of $H_2PO_4^-$ are *acid salts,* because they have fairly readily available hydrogen ions. The calcium salt, $Ca(H_2PO_4)_2$, for example, is used in baking powders as an *acidulant,* a solid that can begin to act as an acid as soon as water is added. In water, the $H_2PO_4^-$ ions delivered by calcium dihydrogen phosphate donate protons to bicarbonate ions, which are present in baking powders as solid $NaHCO_3$. The reaction produces CO_2 and H_2O according to the following net ionic equation.

$(NH_4)_2HPO_4$, diammonium phosphate (DAP), is an important agricultural fertilizer.

$$H_2PO_4^-(aq) + HCO_3^-(aq) \longrightarrow CO_2(g) + H_2O + HPO_4^{2-}(aq)$$

The generation of CO_2 gas within the batter and its expansion during baking create the frothy texture of a baked item.

Polyphosphates Salts that contain the anions of the various polymeric oxoacids of phosphorus are called *polyphosphates.* The anions contain two or more phosphorus atoms per molecule arranged in alternating P—O—P linkages, and such chains can extend to as many as ten phosphorus atoms. In each system the geometry is tetrahedral around P.

We can think of the P—O—P network as forming by the removal of a water molecule between molecules of phosphoric acid. When kept above its melting point, phosphoric acid slowly develops an equilibrium with *diphosphoric acid.*

For $H_4P_2O_7$,

$K_{a_1} = 0.1$
$K_{a_2} = 1.6 \times 10^{-2}$
$K_{a_3} = 2.7 \times 10^{-7}$
$K_{a_4} = 2.4 \times 10^{-10}$

$$\underset{\text{phosphoric acid, } H_3PO_4\text{, two molecules}}{HO{-}\overset{\overset{\displaystyle O}{\|}}{\underset{\underset{\displaystyle OH}{|}}{P}}{-}OH + HO{-}\overset{\overset{\displaystyle O}{\|}}{\underset{\underset{\displaystyle OH}{|}}{P}}{-}OH} \rightleftharpoons \underset{\text{diphosphoric acid, } H_4P_2O_7}{HO{-}\overset{\overset{\displaystyle O}{\|}}{\underset{\underset{\displaystyle OH}{|}}{P}}{-}O{-}\overset{\overset{\displaystyle O}{\|}}{\underset{\underset{\displaystyle OH}{|}}{P}}{-}OH} + H_2O$$

When heated with excess water, the equilibrium in this system shifts to the left, in accordance with Le Châtelier's principle (Sections 11.8 and 15.9). Thus diphosphoric acid hydrolyzes to phosphoric acid. All polyphosphate systems can be similarly hydrolyzed to phosphoric acid (or some mixture of its anions, depending on the pH).

The neutralization of all four available protons of diphosphoric acid gives the diphosphate ion, $P_2O_7^{4-}$, the simplest polyphosphate ion. Its calcium salt, $Ca_2P_2O_7$, is used in toothpastes as an inert, insoluble abrasive that does not interfere with fluoride additives.

The sodium salt of a doubly protonated form of $P_2O_7^{4-}$, $Na_2H_2P_2O_7$, is an acid salt and is also used as an acidulant in baking powders. Its advantage over $Ca(H_2PO_4)_2$, discussed earlier, is that it does not react with bicarbonate ion except when heated. Hence, bakers can prepare large quantities of batter or

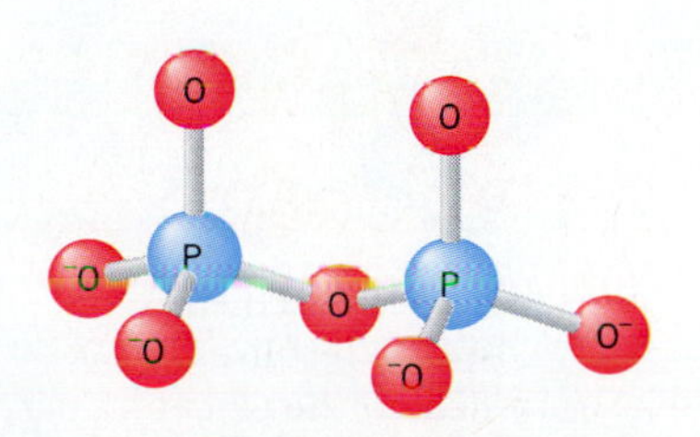

Diphosphate ion, $P_2O_7^{4-}$

dough at room temperature and even store them without the dough-raising action starting prematurely.

Higher members of the polyphosphate family can be thought of as forming by the loss of water between additional molecules of phosphoric acid, much as in the formation of diphosphoric acid. Their structures have the following features, where the subscript *n* indicates the chain length.

$$\mathrm{HO{-}\overset{\overset{\Large O}{\|}}{\underset{\underset{\Large OH}{|}}{P}}\left(-O{-}\overset{\overset{\Large O}{\|}}{\underset{\underset{\Large OH}{|}}{P}}{-}O-\right)_n\overset{\overset{\Large O}{\|}}{\underset{\underset{\Large OH}{|}}{P}}{-}OH}$$

polyphosphoric acids

When $n = 3$, the corresponding compound is *triphosphoric acid,* and like all polyphosphoric acids it can be hydrolyzed to phosphoric acid. Triphosphates in which one hydrogen is replaced by an organic group, as in adenosine triphosphate (ATP), are major "storehouses" of chemical energy in living cells.

Sodium tripolyphosphate is a water-softening agent. The regularly spaced negative charges of its tripolyphosphate ion enable it to form water-soluble complex ions with the metal ions of hard water (Ca^{2+}, Mg^{2+}, and Fe^{3+}) and so prevent the metal ions from forming a precipitate with soap. For example,

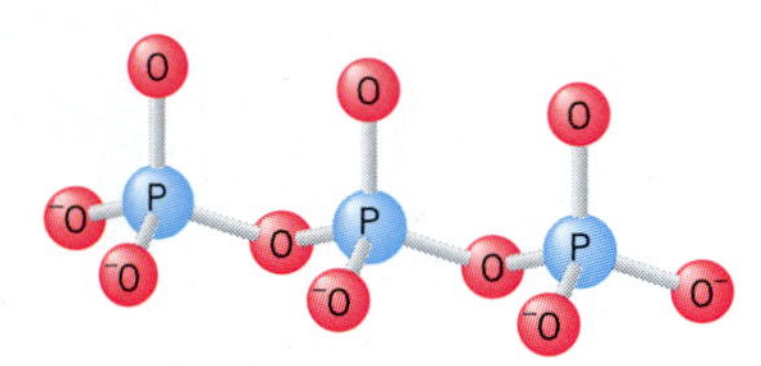

Tripolyphosphate ion, $P_3O_{10}{}^{5-}$

$$\underset{\text{tripolyphosphate ion}}{P_3O_{10}{}^{5-}(aq)} + Ca^{2+}(aq) \longrightarrow \underset{\text{complex between } Ca^{2+} \text{ and tripolyphosphate (water-soluble)}}{CaP_3O_{10}{}^{3-}(aq)}$$

Sodium tripolyphosphate in a laundry product also boosts the cleansing efficiency of the detergent.

Many states have banned the sale of phosphate detergents and Na_3PO_4.

The tripolyphosphate ion was once widely used as the "phosphate" in laundry detergents. The trouble with phosphates in detergents is that they support the life of algae, so when phosphate-containing wastewater gets into rivers, lakes, and bays, they become unfit for aquatic life (except algae) or for human recreation.

Cyclic types of phosphoric acids and their ions are also known. The simplest is *cyclo*-triphosphoric acid, $H_3P_3O_9$, with three phosphorus atoms alternating with oxygen atoms in six-membered rings. (It's commonly called metaphosphoric acid, but the IUPAC name we have used obviously carries more structural information.) Larger rings with up to 10 phosphorus atoms are also known.

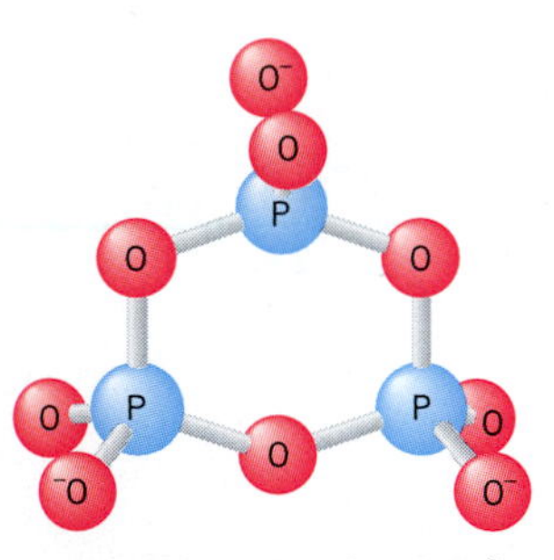

cyclo–Triphosphate ion, $P_3O_9{}^{3-}$

Other Oxoacids of Phosphorus Besides phosphoric acid (in its variety of forms), phosphorus forms two other oxoacids in which the phosphorus occurs in lower oxidation states, namely, phosphorous acid and hypophosphorous acid. Notice the direct bonds between P and H.

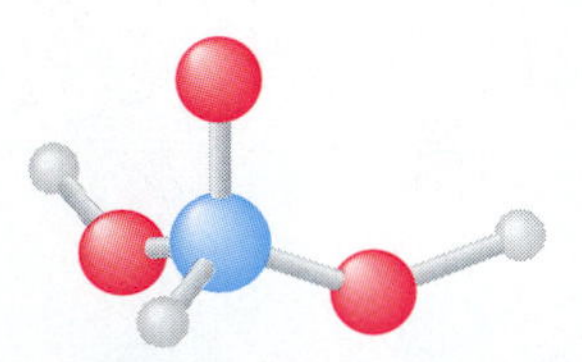

Phosphorous acid H_3PO_3

$$\underset{\text{phosphorous acid}}{\mathrm{HO{-}\overset{\overset{O}{\|}}{\underset{\underset{H}{|}}{P}}{-}OH}} \qquad \underset{\text{hypophosphorous acid}}{\mathrm{H{-}\overset{\overset{O}{\|}}{\underset{\underset{H}{|}}{P}}{-}OH}}$$

Phosphorous acid, H_3PO_3, a colorless solid (MP, 70.1 °C), can be made by the reaction of either P_4O_6 or PCl_3 with water.

$$P_4O_6(s) + 6H_2O \longrightarrow 4H_3PO_3(aq)$$

$$PCl_3(l) + 3H_2O \longrightarrow H_3PO_3(aq) + 3HCl(aq)$$

The H bonded to P in H_3PO_3 cannot be donated as a proton, so phosphorous acid is diprotic, not triprotic. It is a moderately strong acid ($K_{a_1} = 3 \times 10^{-2}$).

The anion of *hypophosphorous acid,* H_3PO_2, forms when white phosphorus reacts with boiling aqueous base.

$$P_4(s) + 4OH^-(aq) + 4H_2O \longrightarrow \underset{\text{hypophosphite ion}}{4H_2PO_2^-(aq)} + 2H_2(g)$$

The free acid, a white crystalline solid obtained by acidifying the mixture, is a moderately strong, monoprotic acid ($pK_a = 1.2$), the two hydrogens covalently bonded directly to phosphorus being unavailable in proton donating reactions.

Antimony (*left*). Arsenic (*right*).

Other Members of Group VA—Arsenic, Antimony, and Bismuth

In the nitrogen family, we move from nonmetals that form acidic oxides—nitrogen and phosphorus—to metalloids that form amphoteric oxides—arsenic and antimony—to the last element—bismuth—that is barely a metal and forms a basic oxide.

Orpiment, As_2S_3

Arsenic, antimony, and bismuth are all brittle, crystalline solids at room temperature, and all conduct electricity poorly. As we go down the family from arsenic through bismuth, the general reactivity of the elements toward other elements and water decreases. All three form oxides of the same two types that we studied for phosphorus—M_4O_6 and M_4O_{10}, where M is the element in question. Arsenic and antimony form acids, H_3MO_4, similar to phosphoric acid. The hydride of arsenic, AsH_3 (arsine), is much more stable than those of antimony and bismuth—SbH_3 (stibine) and BiH_3 (bismuthine)—the latter being extremely unstable. All three hydrides are very toxic.

Realgar, As_4S_4

Trihalides, MX_3, for all four halogens (X) are known. Of the pentahalides, MX_5, only the pentafluorides are known plus the pentachlorides of arsenic and antimony. The oxidation state is +5 in the pentahalides, and the pentafluorides are particularly strong oxidizing agents. BiF_5 is noteworthy; it explodes on contact with water, forming ozone and other products!

The sulfides of arsenic, antimony, and bismuth occur in a variety of ratios of elements and colors. The mineral *orpiment,* As_2S_3, has an almost goldlike color. *Realgar,* As_4S_4, is bright orange-red. *Stibnite,* Sb_2S_3, a black compound, was used in the most ancient of times to darken women's eyebrows. The sulfides are important in schemes for the qualitative analysis of these elements.

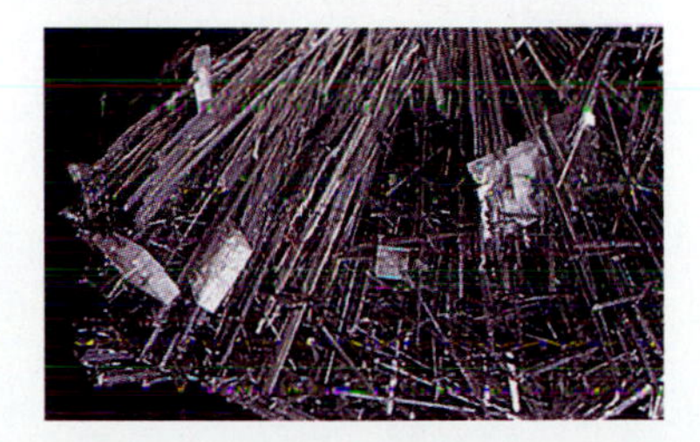

Stibnite, Sb_2S_3

Arsenic forms a compound with gallium, gallium arsenide (GaAs), that has fascinating and useful properties. It is able to convert electricity directly into a laser beam of coherent light and is used in the light-emitting diodes or LEDs of audio disk players and visual display devices. Gallium arsenide is a semiconductor, like silicon, but conducts electricity more rapidly than silicon at the same or lower power, producing less waste heat.

2.3 THE HALOGENS

The **halogen family** consists of fluorine, chlorine, bromine, iodine, and astatine (see Table 2.4). Like the first members of the oxygen and nitrogen families, fluorine is more different from chlorine, the second member, than chlo-

TABLE 2.4 The Halogens

Name	Atomic Symbol	Boiling Point (°C)	Melting Point (°C)	Density
Fluorine	F	−188	−220	1.7 g L^{-1}
Chlorine	Cl	−35	−101	3.2 g L^{-1}
Bromine	Br	59	−7	3.1 g mL^{-1}
Iodine	I	184	114	4.9 g mL^{-1}
Astatine	At	No stable isotope occurs		

Halogen comes from Greek roots meaning "sea-salt producer," after the ability of halogens to combine directly with metals to make salts.

rine differs from the rest of the family. Astatine is intensely radioactive and found only rarely in some uranium deposits. We will not study it further.

Fluorine

Fluorite, also called *fluorospar,* emits light when heated, and so gave its name to the phenomenon of *fluorescence.*

Fluorine, a pale yellow gas, has only one isotope, fluorine-19. It does not occur in the free state but chiefly as the minerals fluorite (CaF_2), cryolite (Na_3AlF_6), and fluorapatite, $[Ca_3(PO_4)_2]_3 \cdot CaF_2$. The exceptional chemical reactivity of fluorine and its high reduction potential long defeated efforts to isolate it from its compounds. Henry Moisson (1858–1907; Nobel Prize, 1906), a French chemist, was the first to obtain fluorine, doing so by electrolyzing KHF_2 dissolved in liquid HF using platinum–iridium alloy electrodes in a platinum vessel. Today fluorine is manufactured from calcium fluoride at a rate of thousands of tons per year.

Although liquid fluorine can be shipped in special casks cooled by liquid nitrogen, most is used where it is made, chiefly to synthesize fluorochemicals. Uranium minerals, for example, are converted into UF_6 in one step in the processing of uranium fuel for nuclear power plants. There is also a demand for special fluorinating agents to make organofluorine compounds, like the well-known nonstick coating, Teflon (polytetrafluoroethylene).

The molecular chains in Teflon are thousands of C atoms long:

```
      F   F   F   F   F
      |   |   |   |   |
etc.—C—C—C—C—C—etc.
      |   |   |   |   |
      F   F   F   F   F
```

Fluorine has the highest reduction potential of all common chemicals, and it reacts violently at room temperature with almost all other elements. Compounds of fluorine can be made with all of the elements except helium, neon, and argon. Water *burns* in fluorine to give HF and oxygen.

$$2F_2(g) + 2H_2O \longrightarrow 4HF(g) + O_2(g)$$

For $F_2(g) + 2e^- \rightleftharpoons 2F^-(aq)$, $E° = +2.87$ V.

Mixed with hydrocarbons, fluorine cracks all carbon–carbon bonds and strips hydrogens, leaving HF and CF_4. You can see why this element must be stored in special metal alloy cylinders.

Fluorine owes its extreme reactivity in part to an unusually low F—F bond energy. Because the fluorine atom is so small, there is strong, destabilizing repulsion between the lone pairs of electrons in the valence shells of F_2. The result is an F—F bond length somewhat longer than expected, so not very much energy is needed to break the bond.

Hydrogen Fluoride and Hydrofluoric Acid Hydrogen fluoride forms explosively when hydrogen and fluorine are mixed, so it is made instead by the action of concentrated sulfuric acid on fluorospar, calcium fluoride.

$$CaF_2(s) + H_2SO_4(l) \longrightarrow 2HF(g) + CaSO_4(s)$$

The boiling point of HF (19.5 °C) is near room temperature, so it is easily condensed. It is marketed as a liquid in special vessels lined with wax or polyethylene or in cylinders made of alloy steel. Glass containers cannot be used because HF dissolves glass! We can write the equation for the reaction by using SiO_2 to represent the chief component of glass.

$$4HF(g) + SiO_2(s) \longrightarrow SiF_4(g) + 2H_2O$$

The SiF_4 can react further:

$$SiF_4(g) + 2HF(g) \longrightarrow H_2SiF_6(l)$$

The action of HF on glass is used to etch designs on glass objects. All but the design is protected by a wax coating, and the unprotected surface areas are attacked by the HF. Some glass dissolves to leave the etched appearance.

The boiling point of HF (19.5 °C) is 104 degrees higher than that of HCl (−84.2 °C) despite having a lower formula mass than HCl. The reason for the higher boiling point is that fluorine is the most electronegative of all elements, so the covalent bond in H—F is polarized much more than the bond in H—Cl. HF molecules, therefore, form hydrogen bonds to each other much more strongly than do molecules of HCl. In the solid state, HF molecules join by hydrogen bonds in extended zigzag chains as follows.

```
          δ−
          F       F       F       F
   ···   / ···   / ···   / ···   /
     H       H       H       H
             δ+
```

hydrogen bonding (···) in HF(*s*)

Hydrogen fluoride dissolves in water to form *hydrofluoric acid.* Its K_a is 6.8×10^{-4}, so it is a weak acid. There is evidence, however, that a high percentage of HF molecules are ionized, but the resulting H_3O^+ and F^- ions are apparently so strongly associated with each other that, effectively, they are not independent. Hydrofluoric acid behaves, therefore, as if it were a weak acid. Despite this, hydrofluoric acid is a dangerous substance. It must be handled with exceptional care.[2]

Another unusual feature of hydrofluoric acid is the presence of the bifluoride ion, HF_2^-, which has no counterpart among the other hydrohalic acids.

$$F^- + \overset{\delta+}{H}—\overset{\delta-}{F} \longrightarrow [F\cdots H\cdots F]^-$$

bifluoride ion

Even though HF and all its salts are poisonous, the fluoride ion in *trace* concentrations in drinking water helps growing bodies develop teeth that are particularly resistant to decay. Tooth enamel is mostly the mineral hydroxyapatite. When fluoride ion is available during the development of enamel, a much harder mineral tends to form instead—fluorapatite.

[2] When spilled on bare skin, hydrogen fluoride causes direct harm and works slowly inward to cause excruciating pain. The cause is largely the ability of the fluoride ion to bind calcium ion as an insoluble compound. The calcium ion is vitally involved in cellular processes and the transmission of nerve signals. The loss of this ion in nerve tissue results in the retention of excess potassium ion to compensate for the loss of positive charge. The excess K^+ stimulates nerve signals and intensifies the pain. Medical attention should be sought even if the initial effects seem to be small. Not only is hydrogen fluoride dangerous, but any compound that can give some of this gas by hydrolysis should be treated very carefully.

FIGURE 2.11
Gaseous chlorine.

$$[Ca_3(PO_4)_2]_3 \cdot Ca(OH)_2 \qquad [Ca_3(PO_4)_2]_3 \cdot CaF_2$$

hydroxyapatite — fluorapatite

Hydroxyapatite contains hydroxide ions, which are more avidly attacked by acids (produced by mouth bacteria feeding on sugars) than is the much weaker Brønsted base, F^-, in fluorapatite. Hence, teeth whose enamel includes fluorapatite resist decay better. To obtain this protection for their children, many communities add fluoride ion to the drinking water at extremely low concentrations, at or below 1 ppm. At higher concentrations, for example, 2 to 3 ppm, fluoride ion can cause mottling of teeth, so careful control of the fluoride ion level in water is important.

Chlorine

Chlorine is a poisonous, yellowish-green gas (see Figure 2.11), named from the Greek *chloros* meaning "yellow-green." It was first isolated in 1774 by Carl Wilhelm Scheele (1742–1786), a Swedish scientist. What Scheele made was not recognized as an *element,* however, until 1810 and the studies by Humphrey Davey (1778–1829), an English chemist.

Isotope	% Abundance
^{35}Cl	75.53
^{37}Cl	24.47

Chlorine consists of two isotopes, namely, chlorine-35 and chlorine-37. It is present in nature mostly as the chloride ion in seawater or in deposits of the minerals halite (NaCl) and sylvite (KCl).

Chlorine ranks in the top 10 of all chemicals manufactured in the United States. Its production in the mid 1990s amounted to over 12 million tons (nearly 150 billion moles), most of it made by the electrolysis of aqueous sodium chloride (see Section 18.4). Chlorine (BP, −10.1 °C) is easily liquefied and shipped under pressure.

About 70% of all chlorine production goes to the manufacture of organochlorine products such as polyvinyl chloride for the manufacture of clear plastic bottles. Another 20% is used to disinfect wastewater and drinking water supplies and to bleach cotton and paper products. (Hydrogen peroxide has outpaced chlorine as the chief bleaching agent for paper and textiles.) The remaining chlorine is used to make a large variety of inorganic compounds needed by the chemicals industry.

"Chlorine" bleach, NaOCl(*aq*), destroys dyes by oxidizing them to colorless compounds.

Hydrogen Chloride and Hydrochloric Acid When high-purity hydrogen chloride is needed, it is made by the direct combination of the elements.

$$H_2(g) + Cl_2(g) \longrightarrow 2HCl(g)$$

A 1 : 1 mole (or volume) mixture of H_2 and Cl_2 is used, and it explodes when exposed to ultraviolet light. Most hydrogen chloride, as much as 90%, however, is obtained as a by-product in the manufacture of organochlorine compounds. It can also be made by heating a mixture of salt (NaCl) and sulfuric acid.

$$NaCl(s) + H_2SO_4(l) \xrightarrow{\text{Heat}} HCl(g) + NaHSO_4(s)$$

Hydrogen chloride boils at − 84.2 °C but is shipped as a liquid under pressure in steel containers. Considerable quantities of hydrogen chloride are used directly, however, to make its aqueous solution, *hydrochloric acid.* The concentrated hydrochloric acid of commerce is 12 *M* and 37% in HCl. Its largest industrial use is to dissolve oxide scale from steel and other metals—a process called "pickling."

Hydrochloric acid is known commercially as *muriatic acid,* and can be purchased under this name in hardware stores.

TABLE 2.5 Oxides of Chlorine

Oxidation State of Chlorine	Formula	Name	Color	Boiling Point (°C)	Melting Point (°C)
+1	Cl_2O	Dichlorine monoxide	Yellow-red	2.2	−116
+4	ClO_2	Chlorine dioxide	Gas: yellow Liquid: red	11	−60
+7	Cl_2O_7	Dichlorine heptoxide	Colorless	80[a]	−96

[a] This is a very explosive liquid.

Hydrochloric acid is the acid of "stomach acid," where it is needed to activate a protein-digesting enzyme. The chloride ion is present in all body fluids.

Oxides and Oxoacids of Chlorine Of the three oxides of chlorine (Table 2.5), the most important is chlorine dioxide, ClO_2. We will briefly describe it plus the four oxoacids of chlorine, $HClO$, $HClO_2$, $HClO_3$, and $HClO_4$.

The formula of hypochlorous acid is written as HClO or HOCl, but the latter is generally preferred.

Oxoacid of Chlorine	K_a
$HOCl$	3.0×10^{-8}
$HClO_2$	$1.1 \times 10^{b-2}$
$HClO_3$	~1
$HClO_4$	Very large

When chlorine is bubbled into water, some of what dissolves reacts to give a mixture of hydrochloric acid and *hypochlorous acid* in the following equilibrium.

$$Cl_2(aq) + H_2O \rightleftharpoons \underset{\text{hydrochloric acid}}{HCl(aq)} + \underset{\text{hypochlorous acid}}{HOCl(aq)}$$

At equilibrium at 25 °C, a saturated solution of chlorine in water has taken up 0.091 mol of Cl_2 per liter. Of this, the molar concentration of unreacted $Cl_2(aq)$ is 0.061 *M*. The concentrations of $HCl(aq)$ and $HOCl(aq)$ are each 0.030 *M*. Thus 67% of the chlorine in saturated chlorine water remains as dissolved Cl_2.

If chlorine is bubbled into an aqueous base, the above equilibrium shifts to consume all of the added Cl_2 as salts containing Cl^- and OCl^- form. Thus, when chlorine is passed into an aqueous slurry of calcium hydroxide, a mixture of calcium chloride and calcium hypochlorite, called "bleaching powder," is made.

The hypochlorite ion can be written as ClO^- or as OCl^-, but the latter is preferred.

$$2Cl_2(aq) + 2Ca(OH)_2(s) \longrightarrow \underbrace{Ca(OCl)_2(s) + CaCl_2(s)}_{\text{components of "bleaching powder"}} + 2H_2O$$

This bleaching agent, being a solid, is easier to handle than gaseous chlorine or aqueous hypochlorous acid. So it is used for general sanitation purposes where special equipment would be inconvenient. But its calcium ion—an ion responsible for water hardness—makes it unattractive for the laundry.

The "liquid bleach" used in households and laundries is a solution of sodium hypochlorite, $NaOCl$, with a pH adjusted to 11 or higher to ensure that the following equilibrium favors OCl^-, not the less stable $HOCl$.

$$HOCl(aq) + OH^-(aq) \rightleftharpoons OCl^-(aq) + H_2O$$

A minor problem with aqueous sodium hypochlorite is that the hypochlorite ion (OCl^-) disproportionates in base. **Disproportionation** is the change of a single substance into two or more products by both an oxidation and a reduction. Thus, when OCl^- ions disproportionate, some are reduced to Cl^- ions and others are oxidized to chlorate ions, ClO_3^-; and the following equilibrium forms.

$$3ClO^-(aq) \rightleftharpoons 2Cl^-(aq) + ClO_3^-(aq) \qquad K_c = 10^{27}$$

Oxidation state of Cl: $+1$ (in ClO^-), -1 (in Cl^-), $+5$ (in ClO_3^-)

The reaction is slow at room temperature despite how strongly the products are thermodynamically favored at equilibrium, giving "liquid bleach" an acceptable shelf life unless it is heated.

The least stable of all of the oxoacids of chlorine is $HClO_2$, *chlorous acid.* It decomposes chiefly by the following reaction when heated, and it cannot be isolated from its aqueous solutions.

$$5HClO_2(aq) \longrightarrow 4ClO_2(aq) + HCl(aq) + 2H_2O$$

With a K_a of 1.1×10^{-2}, chlorous acid is a moderately strong acid.

Salts of chlorous acid are called *chlorites.* Sodium chlorite, $NaClO_2$, is manufactured as a bleaching agent and as an oxidizing agent particularly suited to destroying organic gases of foul odor in certain manufacturing operations.

Chlorine dioxide, ClO_2, is an important bleaching agent used in the manufacture of paper. It is made in the thousands of tons per year by the reduction of $NaClO_3$. Chloride ion is a strong enough reducing agent for this reaction.

$$2ClO_3^-(aq) + 2Cl^-(aq) + 4H^+(aq) \longrightarrow 2ClO_2(aq) + Cl_2(aq) + 2H_2O$$

Generally, dilute aqueous solutions of ClO_2 are used, because this oxide can explode when pure or even in a concentrated solution.

Chloric acid, $HClO_3$, is a strong acid but known only in aqueous solution. Attempts to isolate it by the evaporation of the solvent lead to a complex disproportionation into chlorine, chlorine dioxide, perchloric acid, and oxygen instead.

$$11HClO_3 \longrightarrow 5HClO_4 + 2ClO_2 + 2Cl_2 + 3O_2 + 3H_2O$$

Salts of chloric acid are called *chlorates* and are generally stable compounds. Sodium chlorate ($NaClO_3$) is an example. The chlorates are used as oxidizing agents, bleaches, and in the manufacture of explosives and matches.

Perchloric acid, $HClO_4$, is the only oxoacid of chlorine that is reasonably stable, and then only when it is not made anhydrous. The commercial form is a 70 to 72% solution. Perchloric acid is a very strong acid. When it is concentrated, it is also a *very* strong oxidizing agent, particularly toward almost anything organic—cork, rubber stoppers, filter paper, alcohols—anything that can burn. It can react explosively on contact with these materials.

Salts of perchloric acid, the *perchlorates,* are likewise dangerous in contact with organic matter. Interestingly, in a *dilute* solution, the perchlorate ion, ClO_4^-, is one of the most inert anions known, not even serving as an oxidizing agent.

Bromine

Isotope	% Abundance
^{79}Br	50.54
^{81}Br	49.46

Bromine was first isolated in 1826 by a young French chemist, A. J. Balard (1802–1876), who found that the addition of aqueous chlorine (called "chlorine water") to water from the Montpellier salt marshes in France turned the water deeply yellow. The cause of the color was found to be a liquid element with a terrible smell, which the French Academy named bromine, from *bromos,* Greek for "stench."

Balard's reaction is still the way that bromine is produced from its most common sources, naturally occurring brines. In the reaction, chlorine oxidizes bromide ion (Figure 2.12).

$$Cl_2(aq) + 2Br^-(aq) \longrightarrow 2Cl^-(aq) + Br_2(aq)$$

The bromine is removed by passing either air or steam into the solution, which carries bromine vapors out where they can be condensed and further purified. We can see from the following standard reduction potentials why this reaction goes from left to right, not the reverse.

$$Cl_2(aq) + 2e^- \rightleftharpoons 2Cl^-(aq) \qquad E° = +1.36\ V$$

$$Br_2(aq) + 2e^- \rightleftharpoons 2Br^-(aq) \qquad E° = +1.07\ V$$

Cl_2 has a greater potential for being reduced than Br_2, so its reduction has to be shown as the forward reaction, making the $Br_2/2Br^-$ system run as an oxidation.

Bromine occurs as a nearly 1:1 mixture of two isotopes, bromine-79 and bromine-81. It has a high density, 3.12 g mL^{-1}, and a deep red color (Figure 2.13). Bromine boils at 59.5 °C, low enough to make it quite volatile at room temperature. Not only does bromine have a most irritating odor, its vapors also chemically attack the soft tissue of the nose and throat.[3]

Bromine resembles chlorine in chemical properties, except for reactivity. Bromine combines directly with most elements, but not with as many as chlorine. Oxides and oxoacids of bromine similar to some of those of chlorine are known, but they are unstable and seldom used.

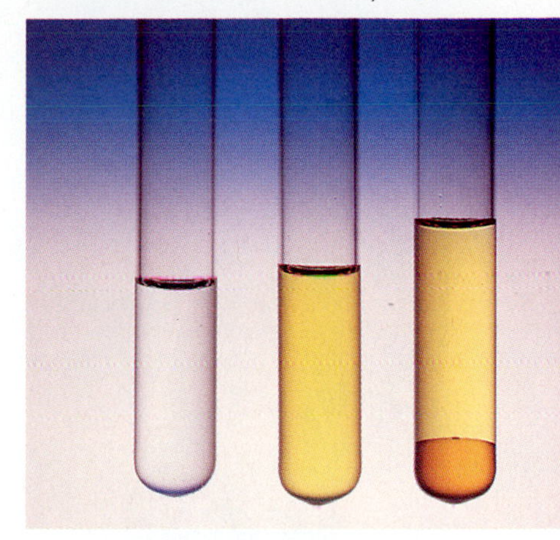

FIGURE 2.12

Bromine forms when the bromide ion is oxidized by chlorine. (*Left*) Aqueous NaBr. (*Center*) $Cl_2(aq)$ has been added to the aqueous NaBr, and the brownish color of Br_2 develops. (*Right*) The newly added globule of carbon tetrachloride, on the bottom, dissolves some of the bromine and so becomes brown.

FIGURE 2.13

Liquid bromine, with a deep brown vapor, is one of only two elements that is a liquid at room temperature, mercury being the other. The vapor pressure of bromine is relatively high, 220 torr, at 25 °C. The liquid is also quite dense, 3.12 g mL^{-1}. These two properties make it difficult to draw liquid bromine into a dropping pipet *and have it stay there* until the bulb is squeezed. It dribbles out, as seen here, and so the transfer of bromine by this method must be done with caution (at a good hood).

[3] Bromine can cause a severe chemical burn on the skin, so this element must always be handled at the hood. Wear easily discarded protective gloves when you pour it. Do not use a medicine dropper; bromine is too dense and will dribble out. It is wise to have some photographers' hypo handy when dispensing bromine. By complex reactions, a paste of the hypo quickly changes bromine to substances less harmful to the skin, at least in the short term. The skin should soon be washed with water, of course, to remove the chemicals produced.

FIGURE 2.14

Solid iodine. Notice the violet color of its vapor.

Bromine is more soluble in water than chlorine but, unlike chlorine, bromine does not react with water to any appreciable extent. A saturated solution of bromine in water at 25 °C is 0.21 *M* in bromine but only 1.2×10^{-3} *M* in *hypobromous acid.* This acid is thus only a very small part of the following equilibrium that is present in "bromine water."

$$Br_2(aq) + H_2O \rightleftharpoons HBr(aq) + HOBr(aq)$$

Hypobromous acid can be prepared by other means, but like hypochlorous acid it is unstable.

The only important acid of bromine is *hydrobromic acid,* $HBr(aq)$, made from hydrogen bromide, $HBr(g)$. Industrially, this gas is made directly by heating the elements in the presence of a platinum catalyst. (Recall that the same kind of reaction between hydrogen and chlorine is explosive and certainly requires no catalyst.)

$$H_2(g) + Br_2(g) \xrightarrow[\text{Pt}]{200-400\ ^\circ\text{C}} 2HBr(g)$$

When hydrogen bromide is bubbled into water, hydrobromic acid results, which is usually sold either at a 50% or a 40% concentration.

Iodine

Only one isotope of iodine occurs naturally.

Iodine was first prepared in 1811 by Bernard Courtois (1777–1838), who observed purple vapors rising from an extract of kelp ashes that he had acidified with sulfuric acid and heated. In his time, kelp was commonly collected, dried, and burned to give ashes from which sodium and potassium salts were obtained. The purple vapor condensed on a cold surface, forming nearly black crystals. Others, notably Joseph Gay-Lussac and Humphrey Davy, proved that the product is an element and named it after the Greek *iodes,* meaning "violet" (Figure 2.14).

Many natural brines contain iodide ion, so when chlorine is bubbled through them, I^- is oxidized to I_2 (Figure 2.15). This reaction is just like the reaction by which bromine is recovered from brines, only the iodide ion is more easily oxidized than the bromide ion, as the standard reduction potentials indicate.

$$Br_2(aq) + 2e^- \rightleftharpoons 2Br^-(aq) \qquad E^\circ = +1.07\ \text{V}$$

$$I_2(s) + 2e^- \rightleftharpoons 2I^-(aq) \qquad E^\circ = +0.54\ \text{V}$$

Bromine, of course, also oxidizes iodide ion, but the calculated cell potential is not as large as with chlorine.

$$2I^-(aq) + Br_2(aq) \longrightarrow I_2(s) + 2Br^-(aq) \qquad E^\circ_{\text{cell}} = 0.53\ \text{V}$$

$$2I^-(aq) + Cl_2(aq) \longrightarrow I_2(s) + 2Cl^-(aq) \qquad E^\circ_{\text{cell}} = 0.82\ \text{V}$$

$$E^\circ_{\text{cell}} = \frac{0.0592}{n} \log K_c$$

These potentials actually correspond to equilibrium constants so high that we are fully justified by writing the equations with single, forward arrows. Using the equation given in the margin, as studied in Section 18.8, the equilibrium constant for the first reaction is 8.0×10^{17} and for the second reaction it is 5.0×10^{27}.

Iodine reacts with far fewer elements or compounds than do the other halogens. Although iodine is a poison, it is not as dangerous as the other members of its family. Its vapors irritate the eyes and mucous membranes.

Iodine kills bacteria, and backpackers have "recipes" for using iodine to disinfect water. *Tincture of iodine,* a solution of iodine in aqueous ethyl alcohol, is used as an antiseptic.

Although I_2 is insoluble in pure water, it dissolves readily in solutions that contain I^- to give I_3^-. The equilibrium is

$$I_2(s) + I^-(aq) \rightleftharpoons I_3^-(aq)$$

Worldwide, hundreds of thousands of tons of iodine are produced annually, and roughly half is used to make various organic iodine compounds.

A solution of hydrogen iodide in water, called *hydriodic acid,* $HI(aq)$, is a strong acid but is little used outside the laboratory. It is made by letting $HI(g)$ pass into water, the latter being made by heating sodium iodide with phosphoric acid.

$$NaI(s) + H_3PO_4(\text{concd}) \longrightarrow HI(g) + NaH_2PO_4(s)$$

(Sulfuric acid is not used because it too easily oxidizes iodide ion back to iodine.)

Hydriodic acid can be purchased as a concentrated solution (40–55% HI). Because oxygen in air, with the aid of light, easily oxidizes the iodide ion to iodine, hydriodic acid is stored in dark brown, well-capped bottles. Even so, the solution slowly darkens as I_2 forms in the following reaction.

$$O_2(g) + 4HI(aq) \longrightarrow 2I_2(s) + 2H_2O$$

Iodine forms a few oxides and oxoacids similar to those of bromine and chlorine, but most are unstable. *Iodic acid,* HIO_3, a white solid, is a strong acid and a strong oxidizing agent. It is the only one of the HXO_3 types of acids (X = Cl, Br, or I) known as a pure compound. Salts of iodic acid, the *iodates,* are commercially available and are used as oxidizing agents in analytical chemistry.

Periodic acid (pronounced PER-i-o-dic), used as an oxidizing agent in organic chemistry, is another crystalline oxoacid of iodine with enough stability to be made and stored. Depending on the degree of hydration, it exists as HIO_4, $H_4I_2O_9$, and H_5IO_6, with the latter being most common.

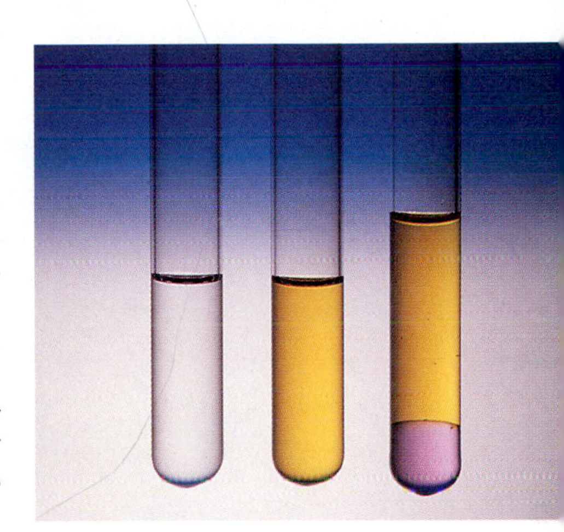

FIGURE 2.15

Iodine forms when the iodide ion is oxidized by chlorine. (*Left*) Aqueous NaI. (*Center*) $Cl_2(aq)$ has been added to the aqueous NaI, and the brownish color that iodine has in water develops. (*Right*) The newly added globule of carbon tetrachloride, on the bottom, dissolves some of the iodine and so takes on the violet color that iodine has in this solvent.

Periodic Correlations among the Halogens

We will draw together here some of the important trends in properties that we have noticed during our study of the halogens and their compounds. We have seen that as we go down the family of the halogens in the periodic table from fluorine to iodine, many properties also diminish in the same order (Table 2.6). As the values of IE, $E°$, and relative electronegativity all show, the first member of the halogen family differs more from the second than the second and remaining members differ from each other.

TABLE 2.6 Trends in the Properties of the Halogens[a]

IE for $X(g) \rightarrow X^+(g) + e^-$ (kJ mol^{-1})	F 1680.6	>	Cl 1255.7	>	Br 1142.7	>	I 1008.7
$E°$ for $X_2(aq) + 2e^- \rightleftharpoons 2X^-(aq)$ (V)	F_2 +2.87	>	Cl_2 +1.36	>	Br_2 +1.07	>	I_2 +0.54
Reactivity toward other elements	F_2	>	Cl_2	>	Br_2	>	I_2
Basicity of X^-	F^-	>	Cl^-	>	Br^-	>	I^-
Relative electronegativity (Pauling scale values)	F 4.1	>	Cl 2.9	>	Br 2.8	>	I 2.2

[a] IE = ionization energy; $E°$ = standard reduction potential.

The order of basicity of X^- ions is just the expected reverse of the order of acidity of H*X*. ("The stronger the acid, the weaker is its conjugate base.")

In Strengths as Acids: HF < HCl < HBr < HI.

Similarly, the order for the values of $E°$, which tells us that F_2 accepts electrons and is far more easily reduced than I_2, means that the order of ease of oxidation of halide ions to halogens, X_2, is the reverse.

This is also the ease with which the ions function as *reducing agents.*

In Ease of Being Oxidized: $I^- > Br^- > Cl^- > F^-$

In these oxidations, the anions are changed to the corresponding elements, and we learned earlier how I^- ions, the most easily oxidized of the halide ions, are oxidized to I_2 molecules by either Br_2 or Cl_2. And we saw how Br^- ions are oxidized to Br_2 molecules by Cl_2. You could not do this oxidation of Br^- using I_2.

We would expect the ionic radii of X^- ions to *increase* as we go down the halogen family, because we go to larger and larger electron clouds.

In Ionic Radii, X^-: $F^- < Cl^- < Br^- < I^-$
in pm: 133 184 196 220

This means that the smallest ion with the highest concentration of negative charge is the fluoride ion, F^-. For this reason, F^- is strongly hydrated in water, as reflected by the order of the enthalpies of hydration of the halide ions. The hydration of F^- is most exothermic.

In Enthalpies of Hydration: $F^- > Cl^- > Br^- > I^-$
kJ mol^{-1}: −515 −381 −347 −305

Finally, one of the most obvious differences we would notice in samples of the halogens is how greatly their colors change; no other family of elements displays such large differences. Fluorine is a pale yellow gas; chlorine is a greenish-yellow gas; bromine is a dark, red-brown liquid; and iodine forms nearly black crystals.

2.4 NOBLE GAS ELEMENTS

According to theory, the helium present when the Earth first formed could not be held by the Earth's gravitational field and was lost into space.

The **noble gases,** Group 0 (Table 2.7), are helium, neon, argon, krypton, xenon, and radon. They are trace components of air and so are isolated by the careful distillation of liquid air. Helium, however, is obtained more economically from natural gas into which it migrated as it formed from the decay of radioactive elements in the Earth's crust. In some geological formations helium is mixed with natural gas at a concentration as high as 7% (v/v). Radon is obtained from the gaseous mixture above aged radium chloride solutions, where it appears as a product of radioactive decay.

TABLE 2.7 The Noble Gas Elements

Element	Symbol	Atomic Number	Boiling Point (°C)	Melting Point (°C)	Color in a Discharge Tube
Helium	He	2	−268.934	−272.2[a]	White to pink-violet
Neon	Ne	10	−246.048	−248.67	Red-orange
Argon	Ar	18	−185.7	−189.2	Violet-purple
Krypton	Kr	36	−107.1	−156.6	Pale violet
Xenon	Xe	54	−61.8	−111.9	Blue to blue-green
Radon	Rn	86		−71	—

[a] Under a pressure of 26 atm.

Mendeleev did not make provision for the noble gases in his periodic table of 1869 because none had been discovered. Evidence that helium is in the sun's atmosphere—helium is named after *helios,* "the sun"—had appeared in late 1868, but only from an otherwise unexplained line in the sun's spectrum. The discovery of argon (1894) created an urgent need for a place in the periodic table, and chemists realized that perhaps an entirely new family of elements existed. All eventually were found. Their family name, once *rare gases,* which they are not, or *inert gases,* which three are not, is now *noble gases.*

***Argon* is from a Greek word meaning "lazy" or "idle."**

Chemical Properties of the Noble Gases

The central fact about the noble gases is their lack of chemical reactivity. They have high ionization energies (see margin table), suggesting a very low potential for entering into any chemical changes that would transfer electrons away or even share them in bonds. Not until we get to xenon do we have a noble gas with a lower first ionization energy than oxygen. In 1933, Linus Pauling predicted that KrF_6 and XeF_6 would be stable enough to exist. In 1962, Neil Bartlett (University of British Columbia) realized that because the first ionization energies of Xe and O_2 are close, xenon should react like oxygen with PtF_6, a reaction that produces $[O_2^+][PtF_6^-]$. The following reaction resulted in the first noble gas compound.

Noble Gas	First IE (kJ mol^{-1})
Helium	2372
Neon	2080
Argon	1520
Krypton	1351
Xenon	1169
Radon	1037
Oxygen, O_2	1170

$$Xe + 2PtF_6 \xrightarrow{25\ ^\circ C} [XeF^+][PtF_6^-] + PtF_5$$

When the mixture of products is warmed to 60 °C, it changes to $[XeF^+][Pt_2F_{11}^-]$. Krypton, xenon, and radon have chemical reactions that lead to relatively stable bonds to fluorine, oxygen, chlorine, and nitrogen.

Xenon, with its larger radius and lower ionization energy, is much more reactive than krypton, and compounds with xenon in oxidation states from +2 to +8 are known. Most involve fluorine or oxygen or both. In addition, metal ions are present in a few xenon compounds.

The three binary fluorides of xenon, XeF_2, XeF_4, and XeF_6, are prepared by heating xenon and fluorine under pressure in special containers. The exact conditions and starting concentrations determine which fluoride forms. All three fluorides are low melting and volatile, suggesting that they are molecular compounds with covalent bonds rather than ionic compounds. XeF_2 is a linear molecule; XeF_4 has a square planar geometry; and XeF_6 has a distorted, nonrigid, octahedral structure (see Section 8.2). All three fluorides of xenon react with water—XeF_4 and XeF_6 violently so. One product is the oxide XeO_3, which itself is dangerously explosive.

All three binary fluorides of xenon are colorless.

The radioactivity of all isotopes of radon has discouraged studies of radon chemistry, but a fluoride is known.

Uses of Noble Gases The relatively higher abundances of helium and argon make these the most used of the noble gases, and virtually all of their applications relate to their lack of chemical reactivity. Thus argon, not air, is used as the low-pressure gas inside a light bulb. Either the oxygen or the nitrogen of air would soon react with the hot metal filament and destroy it, but argon is chemically inert to the glowing metal. Fluorescent bulbs also contain argon instead of air. The familiar "neon" lights (Figure 2.16) contain

Because it is less soluble in blood than nitrogen, helium is a good substitute for N_2 in the "air" made up for deep-sea divers, making it easier for them to avoid the "bends."

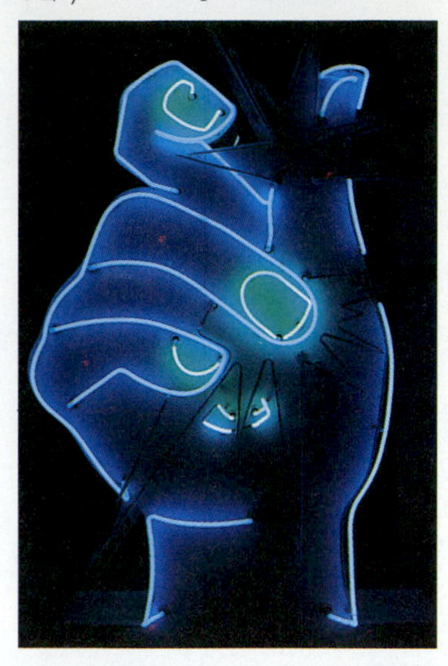

FIGURE 2.16
Different mixtures of noble gases at low pressure in gas discharge tubes make possible the variety of colors of "neon" signs.

mixtures of noble gases whose percentages determine the color seen (see Table 2.7).

Helium as a Superfluid Helium liquefies at 4.18 K (−268.97 °C), and liquid helium is used as a coolant where experiments call for very low temperatures. When liquid helium is cooled to 2.17 K, it undergoes startling changes in physical properties. It loses all resistance to flow—that is, its viscosity is lost—and it is now called a *superfluid.* In this state liquid helium is able to climb the walls of its container!

The degree of coldness made possible by liquid helium causes an interesting change in the ability of some metals, metal alloys, and certain ceramic materials to conduct electricity. When placed in liquid helium, such materials entirely lose their electrical resistance and become *superconductors.* Some of the ceramic substances can even be put into superconducting states at a much higher temperature, at that of liquid nitrogen, which boils at 77.4 K.

Helium is the only substance known that cannot be changed to a solid by cooling it at atmospheric pressure. It takes a pressure of nearly 26 atm to make helium solidify even at absolute zero. It is quite remarkable that there is no combination of temperature and pressure at which dynamic equilibrium exists between all three phases—solid, liquid, and gas—so helium has no triple point (see Section 11.9).

SUMMARY

Sulfur Sulfur occurs as the free element, in sulfide ores, hydrogen sulfide, and sulfate minerals. Of its several allotropes, S_α or orthorhombic sulfur, with crownlike S_8 rings, is the most stable. The same S_8 rings occur in S_β or monoclinic sulfur. Rings and extended chains of sulfur occur in amorphous sulfur. All allotropes eventually revert spontaneously to S_α.

Sulfur forms binary sulfides with hydrogen and most metals and nonmetals. Sulfur oxides, oxoacids, and their salts are particularly important compounds. The only stable hydride is hydrogen sulfide, H_2S.

Binary sulfur–halogen compounds include the unusually stable SF_6 and both S_2Cl_2 and SCl_2, two liquids.

Selenium, tellurium, and polonium form hydrides like H_2S, as well as salts of these hydrides. They also form halides and oxides similar in formula to those of sulfur.

Oxoacids of Sulfur and Their Salts Sulfur burns to give sulfur dioxide, and this forms an aqueous solution called sulfurous acid, commonly written as H_2SO_3 but actually a hydrate of sulfur dioxide, $SO_2 \cdot nH_2O$. Sulfurous acid is a moderate acid that forms solutions of the hydrogen sulfite ion, HSO_3^-, at slightly basic pH values, and solutions of the sulfite ion, SO_3^{2-}, at higher pH values. Commercial "sodium hydrogen sulfite" is not $NaHSO_3$ but consists of sodium disulfite, $Na_2S_2O_5$. However, this provides HSO_3^- ion in solution. Metal sulfites have the sulfite ion in the solid state. The sulfur dioxide/sulfite system is used as a reducing agent.

Sulfur trioxide, made by the catalyzed oxidation of sulfur dioxide, is used to make sulfuric acid, a strong diprotic acid. Hydrogen sulfate salts of the Group IA metals and sulfate salts of most metals are important chemicals. Selenium and tellurium form oxoacids similar to those of sulfur.

Phosphorus Among the many allotropes of phosphorus, white phosphorus (P_4) and red amorphous phosphorus are commercially important. White phosphorus is unstable in air, but its red amorphous allotrope is stable. The binary compound, PH_3 (phosphine), burns in air to give phosphoric acid.

The phosphorus halides react readily with water to give oxoacids—phosphorous acid from phosphorus(III) halides, and phosphoric acid from phosphorus(V) halides. Phosphorus trichloride can be oxidized to phosphorus oxychloride, $POCl_3$, which hydrolyzes to give phosphoric acid.

Phosphorus Oxides, Oxoacids, and Salts P_4O_{10} reacts vigorously with water to give one or a mixture of oxoacids, depending on the relative amount of water. Orthophosphoric acid, H_3PO_4, is the common form referred to when the term "phosphoric acid" is used. Dihydrogen phosphate salts, monohydrogen phosphate salts, and phosphate salts can be made from this moderately strong triprotic acid. Some of the salts are fertilizers; some are acidulants.

Salts of diphosphoric acid, $H_4P_2O_7$, are known, but the most widely used polyphosphate system is found in the tripolyphosphate salts of the $P_3O_{10}^{5-}$ ion of detergents, as well as salts of the *cyclo*-triphosphate system, $P_3O_9^{3-}$.

In its lower oxidation states, phosphorus occurs in phosphorous acid, H_3PO_3, a diprotic acid, and in hypophosphorous acid, H_3PO_2, a monoprotic acid. Reaction of water with P_4O_6 or PCl_3 gives H_3PO_3; reaction of P_4 with boiling base gives the hypophosphite ion, $H_2PO_2^-$.

Arsenic, antimony, and bismuth form the same kinds of oxides (in terms of formulas) as phosphorus, as well as hydrides (arsine, stibine, and bismuthine). Trihalides between these three elements and all four halogens are known, but only the pentafluorides and the pentachlorides of arsenic and antimony are known. Gallium arsenide is a semiconductor.

Halogens The **halogens**—fluorine, chlorine, bromine, and iodine (astatine, being radioactive, is excluded here) —display important similarities, but also significant trends in properties as you go down the halogen family in the periodic table. All halogens form metal halide salts. The oxides and oxoacids of the halogens, when they form, also have similar formulas. Their hydrides are all gases and (except for HF) form strong acids in water. None occurs as the element free in nature but as halide ions in various salts.

In descending the family in the periodic chart, we can say the following.

1. The hydrohalic acids, H*X*(*aq*), become stronger.
2. The standard reduction potentials become smaller.
3. The electronegativity of the element becomes lower.
4. The general reactivity toward anything becomes lower.
5. The basicity of the halide ion becomes weaker.

Fluorine and chlorine are made by electrolysis. Bromine and iodine are made by the oxidation of their halide ions. The hydrogen halides can all be made from metal halides by heating them with sulfuric acid (or phosphoric acid) and driving off the H*X* gas. All H*X* types, except HI, can be made directly from the elements.

Fluorine and water burn. Water and chlorine form an equilibrium with hydrochloric acid and hypochlorous acid. The latter is unstable but its salts are known and have bleaching power. Aqueous bromine contains very little HBr and HOBr. Iodine is quite insoluble in water. Chlorine dioxide, chlorate salts, perchloric acid, and periodic acid are strong oxidizing agents.

Noble Gases The **noble gas elements**—helium, neon, argon, krypton, xenon, and radon—are noteworthy for their lack of chemical reactivity. Krypton forms fluorides and xenon forms several compounds, but mostly with the most electronegative elements—fluorine and oxygen. The most used noble gases are the most abundant, helium and argon. They provide inert blanketings where the exclusion of anything reactive, particularly oxygen, is essential, as inside light bulbs. Mixtures of noble gases are used to make the colors of neon lights.

THINKING IT THROUGH

The goal for each of the following problems is to give you practice in thinking your way through problems. The goal is not *to find the answer itself; instead, you are only asked to assemble the available information or generalizations needed to obtain the answer, state what additional data (if any) are needed, and describe how you would use the information to answer the question.*

1. Suppose that elements *X* and *Z* are in the same chemical family in the periodic table. Which of the two, *X* or *Z*, would have the *higher atomic number* under each of the following conditions?

   (a) As elements, *X* is a stronger oxidizing agent than *Z*.
   (b) The H—*X* covalent bond is less polar than the H—*Z* bond.
   (c) The monatomic anion of *X* is a stronger base than the monatomic anion of *Z*.
   (d) The monatomic anion of *X* is a stronger reducing agent than the monatomic anion of *Z*.

2. If elements *D* and *E* are in the same period of the periodic table, which of the two has the higher atomic number under each of the following conditions?

   (a) The oxide of *E* is acidic but that of *D* is basic.
   (b) The compound of *E* with sulfur gives off a rotten egg odor when stirred with dilute hydrochloric acid, but the oxide of *D* is a gas at room temperature and does not react with HCl(*aq*).
   (c) Element *D* forms no oxides but *E* does. (Neither element is oxygen but both are gases at room temperature.)
   (d) Heating a compound (in air) between *E* and sulfur generates an acrid gas; a compound between *D* and sulfur does not do this.

REVIEW EXERCISES

Answers to questions whose numbers are printed in color are given in Appendix A. The more challenging questions are marked with asterisks.

Sulfur

2.1 In what chemical forms does sulfur occur in nature?

2.2 How is the sulfur in hydrogen sulfide, found in natural gas and oil, converted to elemental sulfur? (Write an equation.)

2.3 What is the name and formula of the most stable sulfur allotrope?

2.4 What is the geometry of the S_8 molecule in orthorhombic sulfur?

2.5 What happens during the conversion of S_α to S_β?

2.6 What is the molecular formula of sulfur in monoclinic sulfur?

2.7 What happens to S_β when it is allowed to stand a long time?

2.8 Why does the viscosity of molten sulfur increase sharply between 160 and 195 °C?

2.9 As the temperature of molten sulfur is increased above 200 °C, the viscosity of the liquid decreases. Explain.

Sulfur Dioxide, Sulfurous Acid, and the Sulfites

2.10 What name is traditionally given to an aqueous solution of sulfur dioxide? Give the names and formulas of the chief chemical forms of the solute in this solution.

2.11 SO_2 has been described as an acidic oxide. Complete and balance the following reactions that illustrate this.
(a) $NaOH(aq) + SO_2(g) \rightarrow$
(b) $NaHCO_3(aq) + SO_2(g) \rightarrow$

2.12 What is the chief sulfur-containing anion in the following?
(a) Solid sodium metabisulfite
(b) Solid "sodium hydrogen sulfite"
(c) Aqueous sodium hydrogen sulfite
(d) Solid sodium bisulfite
(e) Solid sodium sulfite
(f) Aqueous sodium sulfite

2.13 Write the equation for the equilibrium involving sulfur-containing species present in aqueous sodium hydrogen sulfite.

***2.14** Using information in this chapter and in the main text, write the net ionic equation for the reaction in which aqueous sulfur dioxide, $H_2SO_3(aq)$, reacts with iodine to give sodium sulfate and iodide ion. Using standard reduction potentials, calculate the cell potential for this reaction. Will the reaction be spontaneous?

2.15 The addition of dilute hydrochloric acid to a white powder that was known to be either $NaHSO_3$ or $NaHSO_4$ produced a gas with a sharp, choking odor. What is the name and formula of the gas that formed, and what was the formula of the solid? Write molecular and net ionic equations for the reaction.

2.16 Write both the molecular and net ionic equations for the reaction of hydrochloric acid with
(a) $NaHSO_3(aq)$ (b) $Na_2SO_3(aq)$

2.17 Hydrobromic acid, $HBr(aq)$, reacts with hydrogen sulfites and sulfites in the same way as hydrochloric acid (Review Exercise 2.16). Write the molecular and net ionic equations for the reaction of $HBr(aq)$ with each of the following.
(a) $NaHSO_3(aq)$ (c) $K_2SO_3(aq)$
(b) $Na_2SO_3(aq)$ (d) $KHSO_3(aq)$

Sulfur Trioxide, Sulfuric Acid, and the Sulfates

2.18 Complete and balance the following equations that illustrate how sulfur trioxide is an acidic oxide.
(a) $SO_3(g) + H_2O \rightarrow$
(b) $SO_3(g) + NaHCO_3(aq) \rightarrow$
(c) $SO_3(g) + Na_2CO_3(aq) \rightarrow$
(d) $SO_3(g) + NaOH(aq) \rightarrow$

2.19 Write the molecular equations for the individual steps in the *contact process.*

2.20 Why cannot phosphate rock be used *directly* as a source of phosphate for soil enrichment? How does sulfuric acid make this phosphate available?

2.21 Sulfuric acid can just as easily be made as 100% H_2SO_4 as it can as 96% H_2SO_4. Why is sulfuric acid most commonly sold as the 96% form?

2.22 Give five reasons why sulfuric acid is the most frequently used strong acid in industrial applications.

2.23 A solid known to be either $NaHSO_4$ or Na_2SO_4 was dissolved in water and solid Na_2CO_3 was added. The solution fizzed as a gas evolved. What was the gas? What was the original solid? Write a molecular equation for the reaction.

2.24 How would a 1 *M* solution of $NaHSO_4$ test—acidic, basic, or neutral? (If not neutral, write an equation to explain your answer.)

2.25 How would a 1 *M* solution of Na_2SO_4 test—acidic, basic, or neutral? (If not neutral, write an equation to explain your answer.)

2.26 What is Drierite? What is it used for? What property allows for its regeneration?

2.27 How does Na_2SO_4 work as a drying agent?

Other Compounds of Sulfur

2.28 How is sodium thiosulfate made? (Write the molecular equation.)

2.29 The "parent" acid of the thiosulfate ion has what name? Is it stable?

2.30 How does the thiosulfate ion react with (a) Cl_2 and (b) I_2? Write net ionic equations.

2.31 Which compound of sulfur—give its formula and name—has a rotten egg odor?

2.32 A mineral found on a geological survey was suspected of containing either FeS or $FeCO_3$. What simple test could the field geologist perform to find out which it was? (Include an equation.)

2.33 Which is the stronger Brønsted base, S^{2-} or O^{2-}? What fact about H_2S given in this chapter enables one to know?

*2.34 What is the calculated pH of 0.100 *M* NaHS in water at 25 °C?

2.35 Which fluoride of sulfur is exceptionally stable?

2.36 Give the names and formulas of three binary compounds of S and Cl.

2.37 Write the formulas of each.

(a) selenous acid
(b) tellurous acid
(c) selenic acid
(d) telluric acid

2.38 In what structural way does selenium resemble sulfur?

2.39 What are the formulas of the hydrides of selenium and tellurium?

2.40 Complete and balance the following equation, basing your answer on expected similarities to the chemical properties of a corresponding sulfur compound.

$$H_2SeO_4(aq) + 2NaOH(aq) \longrightarrow$$

Phosphorus and Some of Its Binary Compounds

2.41 In terms of bonding abilities of N and P, what might be a reason nitrogen occurs only as N_2, but phosphorus occurs in several allotropes?

2.42 What allotrope of phosphorus forms directly from the carbon-based reduction of calcium phosphate? Write its formula, and describe the structure of its molecules.

2.43 What makes the allotrope of Review Exercise 2.42 so reactive?

2.44 Why do you suppose red amorphous phosphorus is so much less reactive than white phosphorus?

2.45 What is the name and formula of the binary hydride of phosphorus? How does it compare with NH_3 in

(a) Solubility in water?
(b) Basicity?
(c) Ease of oxidation in air?
(d) Odor

2.46 What are the names and formulas of two common binary compounds of phosphorus and chlorine? Write equations for their syntheses.

2.47 Write equations for the reaction of the two binary P—Cl compounds (Review Exercise 2.46) with water, and name the phosphorus-containing products.

2.48 Write equations for the reactions of PBr_3 and PBr_5 with water.

2.49 How is phosphorus oxychloride made? (Write an equation.)

2.50 How does phosphorus oxychloride react with water? Write the equation.

Oxides of Phosphorus and Oxoacids

2.51 How is P_4O_{10} made? (Write an equation.)

2.52 How is H_3PO_4 made (a) from P_4O_{10}? (b) From phosphate rock?

2.53 What is the formula of diphosphoric acid? How is it related to orthophosphoric acid? (Write the equilibrium equation.)

2.54 Sodium dihydrogen phosphate in an *acid salt.* What does this mean?

2.55 In general terms, what does an *acidulant* do in a batter? Also write a net ionic equation.

2.56 Write names for the following compounds.

(a) H_3PO_4
(b) H_3PO_3
(c) H_3AsO_3
(d) KH_2PO_4
(e) $MgHPO_4$
(f) Na_2HPO_4
(g) NaH_2PO_2

2.57 Reasoning by analogy to names and formulas of compounds of phosphorus, what would be the most likely names of the following?

(a) H_3AsO_4
(b) Na_3AsO_4
(c) H_3AsO_3
(d) NaH_2AsO_4
(e) $AsCl_3$
(f) $SbCl_5$

The Halogens and Their Compounds

2.58 Describe how each of the halogens is made from naturally occurring materials. Write equations.

2.59 Write equations that show how each hydrogen halide is made on an industrial scale.

2.60 Write the equations for the reactions (if any) of water with each of the halogens.

2.61 How does hydrogen fluoride dissolve glass? (Write an equation.)

2.62 Why is hydrogen bonding so much stronger in $HF(l)$ than in $HCl(l)$?

2.63 How is bleaching powder made? (Write an equation.)

2.64 Compare and contrast the kinds and relative abundances of various solute species in aqueous chlorine and aqueous bromine. (Use both names and formulas.)

2.65 Write the equations for the disproportionations (a) of OCl^- in base and (b) of ClO_3^- in acid.

2.66 Why does a colorless solution of hydriodic acid darken on standing? What forms to cause the color? Write an equation.

2.67 What property makes perchloric acid particularly dangerous when it is pure?

2.68 *X* and *Y* are both halogens, and the following reaction takes place:

$$X_2 + 2Y^- \longrightarrow 2X^- + Y_2$$

(a) Which halogen is oxidized, *X* or *Y*?
(b) Which halogen is reduced?
(c) If *Y* is Cl, can *X* be Br? Explain.
(d) If *Y* is Br, can *X* be Cl? Explain.
(e) If *Y* is I, can *X* be Br? Explain.

2.69 Arrange the halogens in their correct orders according to the following criteria. Place the symbol of the species that corresponds to the lowest value on the left and the symbol of the species with the highest value on the right in a horizontal row.

(a) the strengths of X_2 as oxidizing agents
(b) the strengths of $HX(aq)$ as acids
(c) the electronegativities of X
(d) the polarities of the bonds in $H—X(g)$
(e) the basicities of $X^-(aq)$
(f) the general reactivities toward metals of X_2
(g) the strengths of X^- as reducing agents

2.70 What chemical property of the iodide ion helps to confirm that phosphoric acid is a weaker oxidizing agent than sulfuric acid?

2.71 What are the names and formulas for two oxoacids of iodine?

Noble Gases

2.72 What are the names and atomic symbols of the members of the noble gas family?

2.73 Which noble gases can form compounds? Give the formulas of some examples.

2.74 Explain why xenon is more reactive than krypton.

2.75 Describe some uses of the noble gas elements.

2.76 Explain why helium and argon are more widely used than the other noble gases.

2.77 Why is the atmosphere in an electric light bulb a noble gas and not air or nitrogen?

2.78 Describe an unusual property of liquid helium.

Additional Problems

***2.79** When white phosphorus and sulfur are heated in the absence of air at a temperature of about 300 °C, a compound of phosphorus and sulfur forms. Its molecular mass was found to be 444. When 0.2141 g of this compound was heated with sufficient 1.0 *M* NaOH to keep the solution always basic, there was obtained a solution to which barium nitrate was added until no more precipitate of barium phosphate formed. There was obtained 0.5798 g of barium phosphate.

(a) What is the empirical formula of the phosphorus–sulfur compound?

(b) What is its molecular formula?

***2.80** "Chile saltpeter," which is a source of sodium nitrate and potassium nitrate, contains some sodium iodate as well. During the isolation of the crystalline nitrates, a brine solution forms that has become concentrated in iodate ion—to a level of 6% iodate. Iodine is isolated from this brine in a two-step process. In the first, sodium hydrogen sulfite solution converts the iodate ion to iodide ion (as the pH of the solution decreases). In the second step, the acidic solution resulting from the first step is mixed with some untreated brine, and the iodate ion and iodide ion react to liberate iodine.

(a) Write a net ionic equation for the first step.

(b) What is the net ionic equation for the second step?

(c) If the initial brine is 6.00% in iodate ion, how many grams of sodium hydrogen sulfite would be needed to change all of the iodate ion in 100.0 g of brine solution into iodide ion?

(d) How many grams of the brine solution would have to be taken for the second step, assuming that 100.0 g of the 6.00% brine solution had been used in the first step?

***2.81** From data supplied in this chapter, write the equation for K_{eq} and calculate the value of K_{eq} for the following reaction.

$$Cl_2(aq) + H_2O \rightleftharpoons \underset{\text{hydrochloric acid}}{HCl(aq)} + \underset{\text{hypochlorous acid}}{HOCl(aq)}$$

Iron is the most widely used structural metal because of its abundance and useful physical properties. Yet the rusting of iron, a chemical property, costs society millions every year. In this chapter we will discuss how metals are obtained from their ores and examine in a systematic way the chemical properties of metals.

Chapter 3

The Properties and Chemistry of Metals

3.1 METALLIC CHARACTER AND THE PERIODIC TABLE

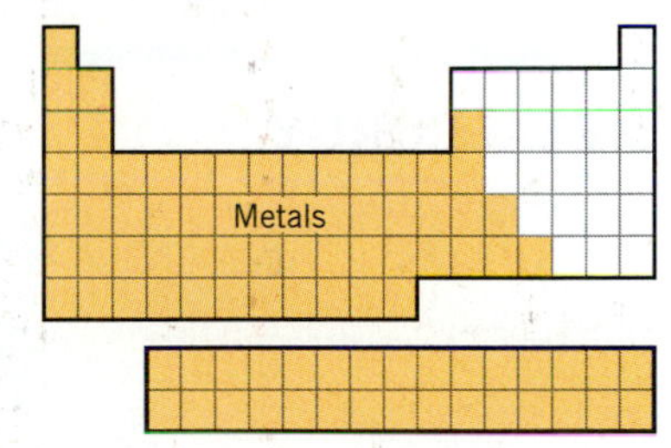

In the preceding two chapters we examined the properties of some nonmetallic elements. Now we turn our attention to the metals to gain a better understanding of their similarities and differences.

The metals make up about 75% of the elements, and we encounter some of them every day. Iron and aluminum, for example, are common structural metals, while others such as copper, silver, gold, and platinum are found in coins and jewelry. Some common metals, such as sodium and calcium, are usually seen only in compounds because of the tendency of the free metals to react with air and moisture. There are also metals that are rare and have only limited applications, so we are hardly aware of their existence or the impact that they may have on our lives, even though some are necessary components of certain enzymes that keep us alive.

The chemical and physical properties of metals, like those of the nonmetals, are determined by the electron configurations of their atoms. Unlike the nonmetallic elements, the atoms of metals in their elemental states have too few electrons to complete their valence shells by forming covalent bonds to other metal atoms. Instead, they achieve electronic stability by forming metallic lattices of the type described in Chapters 8 and 11. In a metallic lattice, you may recall, each metal atom loses valence electrons to the lattice as a whole and all of the metal atoms in the crystal have a share of the pooled electrons. It is the mobility of these electrons that accounts for the unique "metallic luster" and the electrical and thermal conductivity of metals.

You also learned that metals tend to react with nonmetals to form ionic compounds in which the metal exists as a positive ion. The tendency of two elements to form an ionic compound is determined by the difference in the electronegativities of their atoms—the greater the difference, the greater the degree of *ionic character* of the bonds between them. The **ionic character** of a

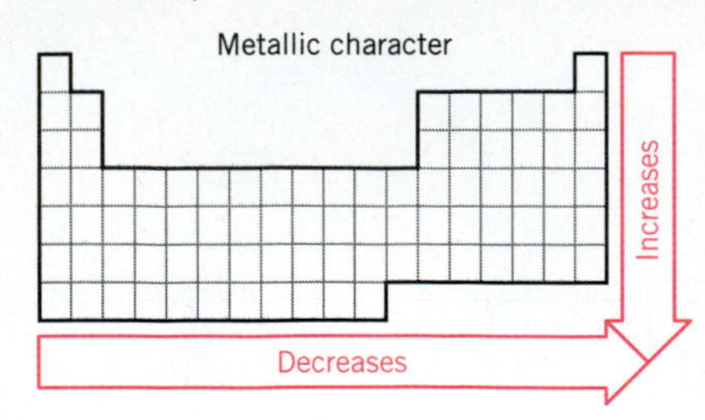

bond can be defined as the extent to which the bond is polar—the more polar the bond the more ionic it is. Because of the way electronegativity varies in the periodic table, increasing from left to right across a period and from bottom to top in a group, the farther away from the lower left corner of the table a metal is located, the more chemically "nonmetallic" it becomes.

Within the periodic table, *metallic character decreases from left to right across a period and increases from top to bottom in a group.* These trends can be observed by noting variations in the acidity or basicity of oxides, because basic oxides are normally associated with metals and acidic oxides with nonmetals. In period 3, for example, sodium and magnesium are metals with typically basic oxides; their oxides react with acids, but not with bases. Aluminum is also a metal, but its oxide is *amphoteric;* it reacts with *both* acids and bases.

$$Al_2O_3(s) + 6H^+(aq) \longrightarrow 2Al^{3+}(aq) + 3H_2O$$

$$Al_2O_3(s) + 2OH^-(aq) \longrightarrow 2AlO_2^-(aq) + H_2O$$

AlO_2^- is just one way to represent the aluminum-containing species that exists in basic solution.

Because Al_2O_3 is able to react with bases, it is at least somewhat acidic. Therefore, we can say that aluminum is "less metallic" than sodium or magnesium. Moving farther to the right in period 3 we come to silicon, which is a metalloid. Then we come to the clearly nonmetallic elements—phosphorus, sulfur, chlorine, and argon.

The increase in metallic character going down a group can be seen most clearly in Group IVA. At the top of the group is carbon, a nonmetal whose oxide, CO_2, dissolves in water to give carbonic acid, H_2CO_3. Next are silicon and germanium, elements with semiconductor properties typical of metalloids. Then come tin and lead, which are metals. Both show some amphoteric properties, but tin more so than lead. Thus tin is less metallic than lead.

3.2 METALLURGY

Metals played an important role in the growth and development of society even before the beginning of recorded history. Archaeological evidence indicates that gold was used in making eating utensils and ornaments as early as 3500 B.C. Silver was discovered at least as early as 2400 B.C., and iron and steel have been used as construction materials since about 1000 B.C. Since these earliest times, the methods for obtaining metals from their naturally occurring deposits have evolved, and today they constitute the subject we call metallurgy.

Metallurgy is the science and technology of metals. In modern terms, it is primarily concerned with the procedures and chemical reactions that are used to separate metals from their ores and the preparation of metals for practical use.

Sources of Metals and Their Compounds

Metals occur in nature in many different ways. Most are found in compounds, either in the Earth's crust or in the oceans, although some of the less reactive metals are found in the uncombined state. A well-known example is gold (Figure 3.1).

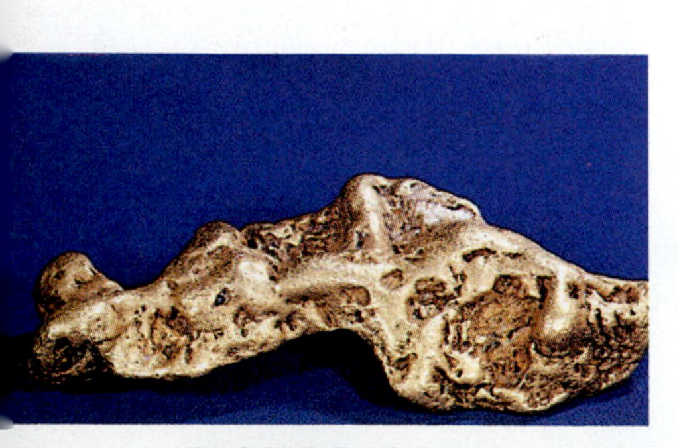

FIGURE 3.1
A gold nugget weighing 82 ounces discovered at Greenville, California.

The distribution of the metals (as well as the nonmetals) throughout the Earth is not uniform, and localized deposits of certain compounds serve as *ores* for various metals. An **ore** is simply a mineral deposit that has a desirable component in a concentration sufficiently high to make its extraction economically profitable. The economics of the recovery operation, though, is what distinguishes an ore from just another rock. For example, magnesium is found in the mineral *olivine,* which has the formula Mg_2SiO_4. This compound contains over 30% magnesium, but there is no economical way to extract the metal from it. Instead, most magnesium is obtained from seawater, even though its concentration is a mere 0.13%. Although the concentration is much less in the water, a profitable method of obtaining the magnesium has been developed, so seawater is the principal source of this metal.

Metals from the Sea The oceans of the world are a storehouse for many different chemicals. These are found not only in the water itself, but also beneath the sea in large deposits. Petroleum, for example, is obtained from many offshore deposits throughout the world. However, you may not be aware of the metals that come from the sea.

The two most abundant cations in seawater are those of sodium and magnesium. Their concentrations are about 0.6 *M* for Na^+ and about 0.06 *M* for Mg^{2+}. Sodium chloride, from which sodium is extracted by electrolysis (Section 18.4[1]), can be obtained in a fairly pure form by the evaporation of seawater in large ponds. About 13% of the salt produced in the United States comes from this source. The rest is mined from vast underground deposits, some of which lie quite deep below the Earth's surface.

Separation of magnesium from seawater takes advantage of the low solubility of magnesium hydroxide. A batch of seawater is made basic by dissolving lime in it. *Lime* is calcium oxide, which is made by decomposing calcium carbonate (limestone), as discussed on page 33. Near the ocean, sea shells, which are composed chiefly of $CaCO_3$, can also provide a cheap source of this raw material.

$$CaCO_3(s) \xrightarrow{\text{heat}} CaO(s) + CO_2(g)$$

In seawater, lime reacts to form calcium hydroxide.

$$CaO(s) + H_2O \longrightarrow Ca^{2+}(aq) + 2OH^-(aq)$$

The hydroxide ion precipitates insoluble magnesium hydroxide, $Mg(OH)_2$.

$$Mg^{2+}(aq) + 2OH^-(aq) \longrightarrow Mg(OH)_2(s)$$

This precipitate is then filtered from the seawater and dissolved in hydrochloric acid.

$$Mg(OH)_2(s) + 2H^+(aq) + 2Cl^-(aq) \longrightarrow Mg^{2+}(aq) + 2Cl^-(aq) + 2H_2O$$

Evaporation of the resulting solution yields magnesium chloride, which is dried, melted, and finally electrolyzed to give free magnesium metal.

$$MgCl_2(l) \xrightarrow{\text{electrolysis}} Mg(l) + Cl_2(g)$$

Another potential source of metals from the sea is the mining of manganese nodules from the ocean floor. **Manganese nodules** are lumps about the size of an orange that contain significant amounts of manganese (almost 25%) and iron (about 15%). In some places, they occur in large numbers, as Figure 3.2 shows. Although the procedures and equipment needed to recover these nodules from the ocean floor are very expensive, the huge number of nodules that

FIGURE 3.2
A large pile of manganese nodules on the ocean floor.

[1] Unless noted otherwise, cross-references to Sections, Special Topics, and *Chemicals in Use* refer to the accompanying text, *Chemistry: The Study of Matter and Its Changes,* 2nd edition (1996) by Brady and Holum, not to this supplement. Some material from text Chapters 15 to 19 is assumed for this supplement.

is believed to exist makes the concept of deep-ocean mining attractive, more so as land-based sources of these metals become depleted.

Metals from the Earth We have turned to the oceans as a source of metals only in recent times. Land-based ores provided civilizations with their earliest access to metals, and even today most of our metals come from ores mined from the Earth.

When the source of a metal is an ore that is dug from the ground, considerable amounts of sand and dirt are usually mixed in with the ore. To reduce the volume of material that must be processed, the ore is normally *concentrated* before the metal is separated from it. How this is done depends on the physical and chemical properties of the ore itself, as well as those of the impurities.

Gangue is pronounced "gang."

In some cases, the unwanted rock and sand, called **gangue,** can be removed simply by washing the material with a stream of water. This flushes away the waste and leaves the enriched ore behind. Some iron ores are treated in this way. This procedure also forms the basis for the well-known technique called "panning for gold" that you've probably seen in movies. A sample of sand that might also contain gold is placed into a shallow pan partially filled with water. As the water is swirled around, it washes the less dense sand over the rim of the pan, but leaves any of the more dense bits of gold behind.

Panning for gold is an activity many tourists practice when they visit Alaska.

Flotation is a method commonly used to enrich the sulfide ores of copper and lead. The ore is crushed, mixed with water, and ground into a souplike slurry which is then transferred to flotation tanks (Figure 3.3) where it is mixed with detergents and oil. The oil adheres to the particles of sulfide ore, but not to the particles of sand and dirt. Air is blown through the mixture and the rising air bubbles become attached to the oil-coated ore particles, bringing them to the surface where they are held in a froth. The detergents in the mixture stabilize the bubbles long enough for the froth and its load of ore particles to be skimmed off. Meanwhile, the sand and dirt settle to the bottom of the tanks and are removed.

Many ores must undergo a second round of pretreatment before the metal can be obtained from them. For example, after enrichment, sulfide ores are

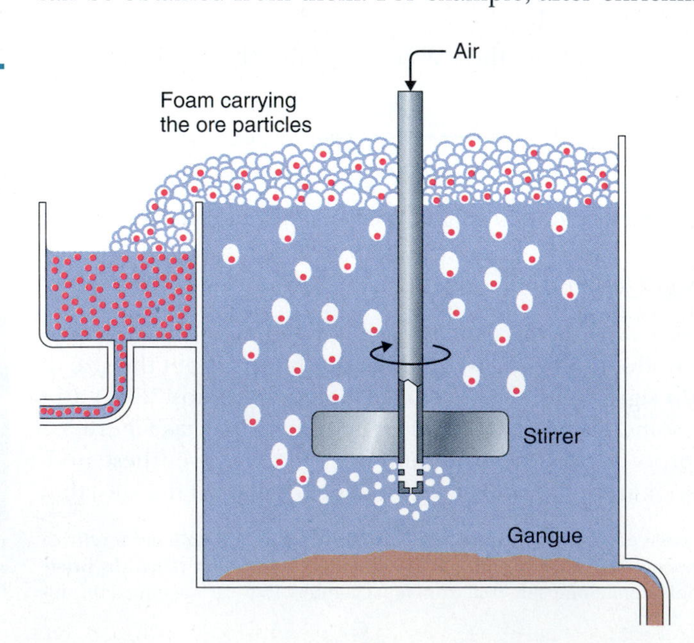

FIGURE 3.3
A flotation apparatus.

usually heated in air. This procedure, called **roasting,** converts the sulfides to oxides, which are more conveniently reduced than sulfides. Typical reactions that occur during roasting are

$$2Cu_2S + 3O_2 \longrightarrow 2Cu_2O + 2SO_2$$

$$2PbS + 3O_2 \longrightarrow 2PbO + 2SO_2$$

A by-product of roasting is sulfur dioxide. In years past this was simply released to the atmosphere and was a major source of air pollution. Figure 3.4, for example, shows the effect on the environment of the processing of copper ore at Copper Hill, Tennessee. Today we realize that the SO_2 cannot be allowed to escape into the air. One way to remove it from the exhaust gases is to allow it to react with calcium carbonate ($CaCO_3$).

$$CaCO_3(s) + SO_2(g) \longrightarrow CaSO_3(s) + CO_2(g)$$

This method creates another problem, however—the disposal of the solid calcium sulfite. Another way to dispose of the SO_2 is to oxidize it to SO_3. The SO_3 can be converted to sulfuric acid, which is then sold.

Aluminum ore, called *bauxite,* must also be pretreated before it can be processed. Bauxite contains aluminum oxide, Al_2O_3, but a number of impurities are also present. To remove the impurities, use is made of aluminum oxide's amphoteric behavior. The ore is mixed with a concentrated sodium hydroxide solution, which dissolves the Al_2O_3.

Aluminum oxide, Al_2O_3, is also called alumina.

$$Al_2O_3(s) + 2OH^-(aq) \longrightarrow 2AlO_2^-(aq) + H_2O$$

The major impurities, however, are insoluble in base, so when the mixture is filtered, the impurities remain on the filter while the aluminum-containing solution passes through. The solution is then neutralized with acid, which precipitates aluminum hydroxide.

$$AlO_2^-(aq) + H^+(aq) + H_2O \longrightarrow Al(OH)_3(s)$$

When the precipitate is heated, water is driven off and the oxide is formed.

$$2Al(OH)_3(s) \xrightarrow{\text{heat}} Al_2O_3(s) + 3H_2O(g)$$

This purified Al_2O_3 then becomes the raw material for the Hall–Héroult process discussed in Section 18.4.

FIGURE 3.4

The destructive effects of SO_2 pollution on the environment around Copperhill, Tennessee are so widespread they can be seen from space, as revealed by this satellite photo.

Separating Metals from Their Compounds

In general, producing a free metal from one of its compounds involves reduction. This is because, with very few exceptions, metals in compounds have positive oxidation states. They must therefore "gain electrons" to become free elements. The nature of the reducing agent needed to provide these electrons depends on how difficult the reduction process is. If the free metal itself is very reactive, only electrolysis can provide enough energy to cause the decomposition of its compounds. This is the reason active metals such as sodium, magnesium, and aluminum are produced electrolytically. Less active metals such as lead or copper require less active reducing agents to displace them from their compounds, and chemical reducing agents can do the job.

A plentiful, and therefore inexpensive, reducing agent used to produce a number of metals is carbon, which is usually obtained from coal. When coal is heated strongly in the absence of air, volatile components are driven off and **coke** is formed. Coke is composed almost entirely of carbon. Carbon is an

Heating an ore with a reducing agent is called **smelting.**

effective reducing agent for metal oxides because it combines with the oxygen to form carbon dioxide. For example, after it is roasted, lead oxide is mixed with coke and heated.

$$2PbO(s) + C(s) \xrightarrow{\text{heat}} 2Pb(l) + CO_2(g)$$

The high thermodynamic stability of CO_2 serves as one driving force for this reaction. The loss of CO_2 and the inability to reach an equilibrium is another.

Copper oxide ores can also be reduced with carbon.

$$2CuO + C \longrightarrow 2Cu + CO_2$$

This step is unnecessary for some copper sulfide ores if the conditions under which the ore is roasted are properly controlled. For example, heating an ore that contains Cu_2S in air can convert some of the Cu_2S to Cu_2O.

$$2Cu_2S + 3O_2 \longrightarrow 2Cu_2O + 2SO_2$$

At the appropriate time the supply of oxygen is cut off and the mixture of Cu_2S and Cu_2O reacts further to give metallic copper.

$$Cu_2S + 2Cu_2O \longrightarrow 6Cu + SO_2$$

Coke is made in a battery of side-by-side coke ovens where coal is heated to drive off volatile materials. Here we see a fresh batch of white-hot coke being pushed from one of the ovens into a waiting railroad car. It will be delivered to a blast furnace where it will be used to reduce iron ore to metallic iron.

Reduction of Iron Ore Without question, the most important use of carbon as a reducing agent is in the production of iron and steel. The chemical reactions take place in a huge tower called a **blast furnace** (see Figure 3.5). Some are as tall as a 15-story building and produce up to 2400 tons of iron a day. They are designed for continuous operation, so the raw materials can be added at the top and molten iron can be tapped off at the bottom. Once started, a typical blast furnace may run continuously for 2 years or longer before it is worn out and must be rebuilt.

To make 1 ton of iron requires 1.75 tons of ore, 0.75 ton of coke, and 0.25 ton of limestone.

The material put into the top of the blast furnace is called the **charge.** It consists of a mixture of iron ore, limestone, and coke. A typical iron ore consists of an iron oxide (Fe_2O_3, for example) plus impurities of sand and rock. The coke is added to reduce the iron oxide to the free metal. The limestone is added to react with the high-melting impurities to form a **slag,** which has a lower melting point. The slag can then be drained off as a liquid at the base of the furnace.

This blast of hot air is what gives the blast furnace its name.

To understand what happens in the furnace, it is best to begin with the reactions that take place near the bottom. Here, heated air is blown into the furnace, where carbon (from the coke) reacts with oxygen to form carbon dioxide.

$$C + O_2 \longrightarrow CO_2$$

The reaction is very exothermic, and the temperature in this part of the furnace rises to nearly 2000 °C. It is the hottest region of the furnace. The hot CO_2 rises and reacts with additional carbon to form carbon monoxide.

$$CO_2 + C \longrightarrow 2CO$$

The active reducing agent in the blast furnace is carbon monoxide.

This reaction is endothermic, which causes the temperature higher up in the furnace to drop to about 1300 °C. As the carbon monoxide rises through the charge, it reacts with the iron oxides and reduces them to the free metal. The reactions are

$$3Fe_2O_3(s) + CO(g) \longrightarrow 2Fe_3O_4(s) + CO_2(g)$$
$$Fe_3O_4(s) + CO(g) \longrightarrow 3FeO(s) + CO_2(g)$$
$$FeO(s) + CO(g) \longrightarrow Fe(l) + CO_2(g)$$

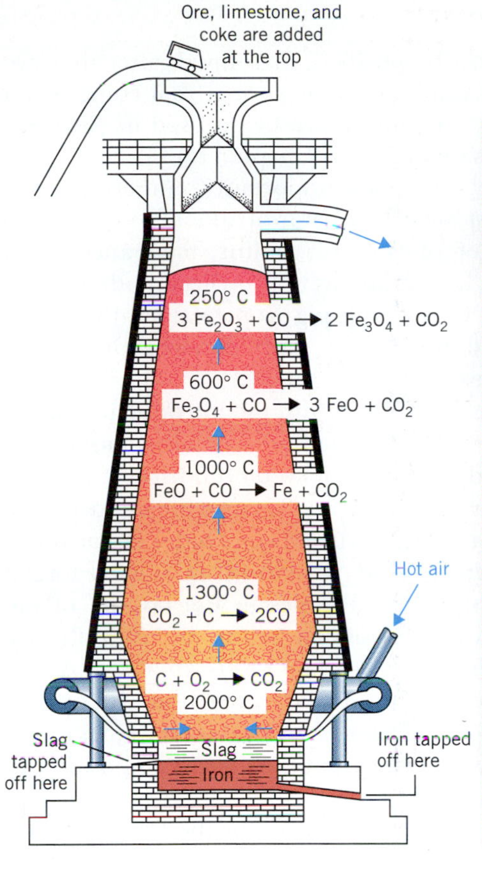

FIGURE 3.5

A typical blast furnace for the reduction of iron ore.

As the charge settles toward the bottom, molten iron trickles down and collects in a well at the base of the furnace.

The high temperature in the furnace also causes the limestone in the reaction mixture to decompose to give calcium oxide.

$$\underset{\text{limestone}}{CaCO_3} \longrightarrow CaO + CO_2$$

The calcium oxide reacts with impurities such as silica (SiO_2) in the sand to form the slag.

The acidic anhydride, SiO_2, reacts with oxide ion in CaO to form the silicate ion, SiO_3^{2-}, in $CaSiO_3$.

$$CaO + SiO_2 \longrightarrow \underset{\text{calcium silicate (slag)}}{CaSiO_3}$$

The molten slag also trickles down through the charge. It collects as a liquid layer on top of the more dense molten iron. Periodically, the furnace is tapped and the iron and slag are drawn off. The iron, which still contains some impurities, is called **pig iron.**[2] It is usually treated further to produce steel. The slag itself is a valuable by-product. It is used to make insulating materials and is one of the chief ingredients in the manufacture of portland cement.

[2] The name pig iron comes from an early method of casting the molten iron into bars for shipment. The molten metal was run through a channel that fed into sand molds. The arrangement looked a little like a litter of pigs feeding from their mother.

Preparing Metals for Use

Before metals can be used, most must be purified, or **refined,** after they are reduced to the metallic state. For example, metallic copper that comes from the smelting process is about 99% pure, but before being used in electrical wiring it is purified electrolytically as described in Chapter 18.

The conversion of pig iron to steel is the most important commercial refining process. The pig iron from a blast furnace consists of about 95% iron, 3 to 4% carbon, and smaller amounts of phosphorus, sulfur, manganese, and other elements. Steel contains much less carbon as well as certain other ingredients in very definite proportions. Converting pig iron to steel, therefore, involves removing the impurities and much of the carbon, and adding other metals in precisely controlled amounts.

The production of steel on a large scale began with the introduction of the Bessemer converter in England in 1856. In this method, a batch of molten pig iron weighing about 25 tons is placed in a large tapered cylindrical vessel that is lined with a *refractory* (very high melting) material. Because the impurities in the iron are usually silicon, sulfur, or phosphorus, whose oxides are acidic, a basic lining of CaO or MgO is normally used. A blast of air is then forced through the molten metal from a set of small holes in the bottom of the container. The oxygen in the air oxidizes the silicon, sulfur, and phosphorus, and the oxides that form react with the CaO or MgO to form a slag. At the same time, the carbon content of the steel is reduced by reaction with oxygen to form CO or CO_2. The conversion of pig iron to steel by this method is rapid, giving an impressive display of fire and sparks, but the exact composition of the finished product is difficult to control.

Recall that metal oxides are basic anhydrides.

Because of the variable quality of the steel produced by the Bessemer process, it was largely replaced by a more easily controlled method that makes use of an **open hearth furnace,** so called because it contains a large shallow hearth (floor) into which the charge is placed. The floor is usually lined with a basic refractory such as CaO or MgO, and the charge consists of a mixture of molten pig iron, Fe_2O_3, scrap iron, and limestone. A blast of hot air and burning fuel is played across the surface of the charge to keep it in a molten state (Figure 3.6). The Fe_2O_3 gradually reacts with the carbon in the mixture, forming bubbles of CO_2 that rise to the surface and keep the mixture stirred. The limestone decomposes and the resulting CaO reacts with impurities to form a slag. Other metals, such as manganese or chromium, are also sometimes added to give the steel special properties such as hardness or resistance to corrosion. Finally, after 8 to 10 hr, the steel is ready to be fabricated into the various forms required by steel users.

A molten iron charge is added to a basic oxygen furnace.

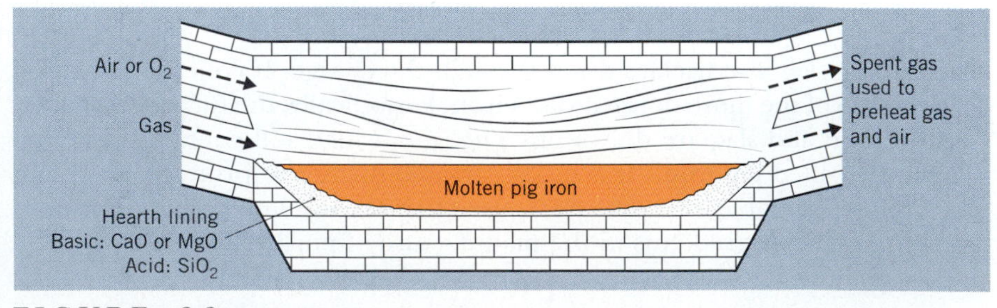

FIGURE 3.6

Cross section of an open hearth furnace.

Although the open hearth furnace produces steel of reproducibly high quality, it is slow (and therefore expensive). Fast computers teamed with modern, high-speed methods of analysis have permitted the steel industry to return to a modification of the older Bessemer process. This modern method is called the **basic oxygen process.** It uses a large pear-shaped reaction vessel that is mounted on pivots and is lined with an insulating layer of special bricks, as shown in Figure 3.7. The charge consists of about 30% scrap iron and scrap steel and about 70% molten pig iron, which melts the scrap. A tube called an *oxygen lance* is then dipped into the charge and pure oxygen is blown through the molten metal. The oxygen rapidly burns off the excess carbon and oxidizes impurities to their oxides. These form a slag with calcium oxide that comes from powdered limestone, which is also added. Finally, other metals are introduced in the proper proportions to give a product with the desired properties. After the steel is ready, the reaction vessel can be tipped to pour out its contents. This method of making steel is very fast. A batch of steel weighing 300 tons can be made in less than an hour.

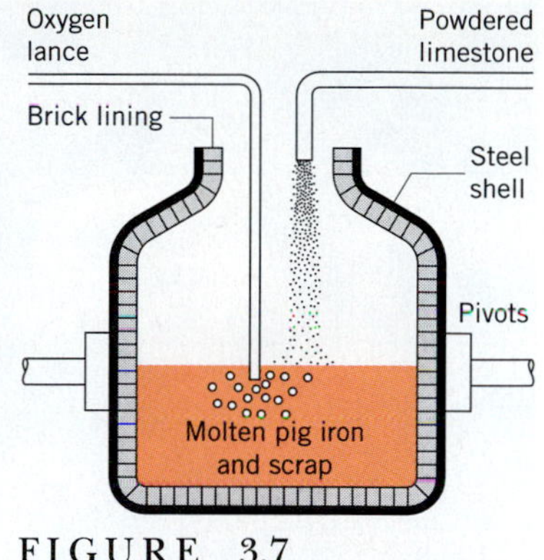

FIGURE 3.7
Diagram of a basic oxygen furnace used for the production of steel.

3.3 THE METALS OF GROUP IA: THE ALKALI METALS

The properties of many metal compounds have been discussed in our study of the chemistry of the nonmetals in Chapters 1 and 2 of this supplement.

In the previous section we discussed how metals are obtained from their ores. One of our goals was to understand how the methods used to do this depend on the specific chemical properties of the metals, as well as on the properties of the ores in which the metals are found. Now we refocus our discussion around the periodic table and consider the chemical and physical properties of the metals in relationship to the groups in which they occur. When we do this, we find the strongest group similarities among the representative metals.

The Metals of Group IA

The elements of Group IA are hydrogen, lithium, sodium, potassium, rubidium, cesium, and francium. Hydrogen, at the top of the group, is a nonmetal, and its chemistry was discussed in Chapter 1 of this supplement. As you saw, hydrogen doesn't really fit well in any group. It owes its location in Group IA to its electron configuration, $1s^1$, rather than to its chemical properties. All the rest of the elements in Group IA are metals, however, and they are generally known by their group name—the **alkali metals.**

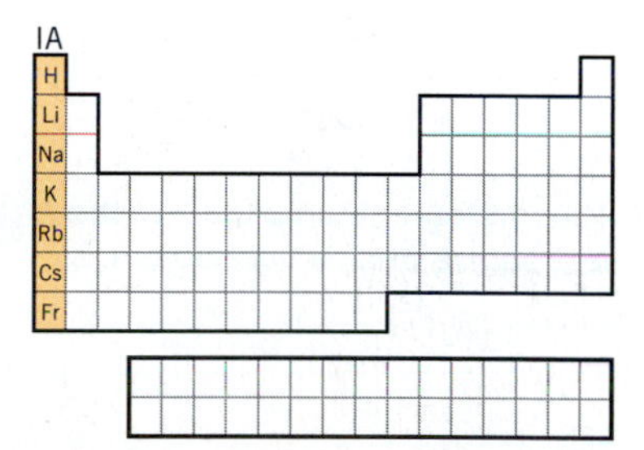

Occurrence

All the alkali metals are so easily oxidized by oxygen and by water that they never are found as free elements in nature; they always occur in compounds. And because nearly all their salts are soluble in water, they are often present as ions in aqueous solution, primarily in seawater, in deep brine wells, and in certain lakes such as the Great Salt Lake in Utah and the Dead Sea on the border between Israel and Jordan. Compounds of the alkali metals are also found in mineral deposits on and below the Earth's surface. For example, large underground deposits of solid salt (NaCl) are common, and there are also surface deposits in some dry salt lakes such as the Bonneville Salt Flats in Utah. Large deposits of sodium sulfate are found in North Dakota, Canada, and in the southwestern United States. In the desert regions near the border between Chile and Peru there are vast deposits of sodium nitrate, a compound

Sodium chloride occurs in large deposits both below and above ground. (*Left*) The interior of a salt mine located in Texas. (*Right*) The Bonneville Salt Flats in Utah.

Saltpeter is KNO_3; Chile saltpeter is $NaNO_3$.

often referred to in commerce as *Chile saltpeter*. In addition, some clays contain alkali metal ions along with aluminum, silicon, and oxygen.

By far, the most abundant of the alkali metals are sodium and potassium. By mass, they rank sixth and seventh among the elements in the Earth's crust. It is perhaps no surprise, therefore, that they are also the most biologically important elements in Group IA. Both sodium and potassium ions are important for animals, but potassium ions are far more important than sodium ions in plants. In fact, with few exceptions (seaweed, for example), sodium ions tend to be toxic to plants.

Lithium, which is relatively rare, has shown promise in the treatment of mental disorders. Lithium carbonate, for example, is used in the treatment of manic depression.

Francium's longest lived isotope is ^{223}Fr, which has a half-life of only 22 min.

Both rubidium and cesium are rare and have little commercial importance. Francium has only a fleeting existence because all its isotopes are radioactive and have very short half-lives.

Physical Properties

The photograph on page 2 of the text shows freshly cut sodium.

Some of the physical properties of the alkali metals are summarized in Table 3.1. All the elements are typically metallic with a bright luster and high thermal and electrical conductivity. The metals are also soft and have low melting points. Their atoms, each with only one valence electron, exist as cations with a charge of 1+ in their metallic lattices. These ions are attracted only weakly to

TABLE 3.1 Some Properties of the Alkali Metals

Element	Electron Configuration	Ionization Energy (kJ/mol)	Density (g/cm^3)	Melting Point (°C)	Reduction Potential (V)[a]
Lithium	[He] $2s^1$	520.1	0.534	180.5	−3.05
Sodium	[Ne] $3s^1$	495.1	0.971	97.8	−2.71
Potassium	[Ar] $4s^1$	418.7	0.862	63.7	−2.92
Rubidium	[Kr] $5s^1$	402.9	1.53	39.0	−2.93
Cesium	[Xe] $6s^1$	375.6	1.83	28.6	−2.92
Francium	[Rn] $7s^1$	—	—	—	—

[a] For $M^+(aq) + e^- \rightarrow M(s)$.

the surrounding electron sea, so the lattice is easily deformed (hence, soft) and little thermal energy is needed to cause the solid to melt. The alkali metals also have low densities because they have the largest atoms in their respective periods; large atoms lead to small ratios of mass to volume.

The alkali metals have low ionization energies. The single electron outside a noble gas core is easy to remove, so the metals are easily oxidized. (We can also say they are good reducing agents.) Because they are so easily oxidized, the free elements have few practical uses as metals, and none that allows them to be exposed to air or moisture. However, sodium has shown promise as a coolant in certain kinds of nuclear reactors. Sodium has a low melting point and a reasonably high boiling point, and it conducts heat well. It can be pumped through the core of the reactor, where it readily picks up heat, and then through tubes in a heat exchanger, where the heat is removed.

Nuclear reactors are described in Chapter 20.

Because the alkali metals are easily oxidized, their cations are difficult to reduce, which is reflected in their very negative reduction potentials. But here we observe a curious trend. Within a group the ionization energy decreases from top to bottom. This suggests that it should become progressively easier to oxidize the metal, and progressively more difficult to reduce the M^+ cation, going from top to bottom. We do observe this from Na to Rb, but lithium is out of line. It has the most negative reduction potential of all. The reason is that reduction potentials deal with the reduction of the metal ion in *aqueous solution.*

$$M^+(aq) + e^- \longrightarrow M(s)$$

Anything that helps to stabilize the metal ion should make it more difficult to reduce, and it is here that lithium ion has a special advantage. The Li^+ ion is very small, so its positive charge is highly concentrated. Also, it is able to get very close to the water molecules that surround it, so it attracts them very strongly. Quantitatively, this is expressed in terms of the ion's hydration energy—the energy that would be released when the cation is placed into its surrounding cage of water molecules. Because of its small size, the lithium ion has a very large hydration energy that must be overcome when the cation is reduced, so Li^+ is very difficult to reduce. The other cations—Na^+, K^+, etc.—are considerably larger, and they have smaller hydration energies that don't differ very much from one to another. They are easier to reduce than Li^+, and the group trend in ionization energy, which is more significant than any differences in the hydration energies of the ions, gives the expected variation in $E°$ as we descend the rest of the group.

Preparation of the Alkali Metals

Because the alkali metals are so easily oxidized, their compounds are very difficult to reduce. Therefore, very few chemical reducing agents are able to do the job. For this reason, the electrons for the reduction are usually supplied by electrolysis, especially for lithium and sodium. In carrying out the electrolysis, a molten salt must be used rather than an aqueous solution because water is more easily reduced than the metal ions. For instance, sodium and lithium are usually prepared from their molten chlorides, NaCl and LiCl. Chlorides are used because their melting points tend to be lower than those of other compounds such as the oxides.

Interestingly, potassium can be made by the reaction of potassium chloride with metallic sodium, even though potassium ions are more difficult to reduce than sodium ions. This is done by passing sodium vapor over molten potassium chloride. The equilibrium

$$KCl(l) + Na(g) \rightleftharpoons NaCl(l) + K(g)$$

is shifted to the right because potassium is more volatile than sodium. The sodium condenses while the potassium is vaporized and swept away as it forms. Metallic rubidium and cesium are prepared in the same way.

Compounds and Reactions

Reduction of H_2O produces H_2; oxidation of H_2O gives O_2.

The Group IA metals are the most reactive of all the metals. They are such powerful reducing agents that they readily reduce water. When placed in water, they react vigorously, releasing hydrogen and forming the metal hydroxide. (The reaction of sodium with water is shown in the photograph on page 384 of the text.)

$$2M(s) + 2H_2O \longrightarrow 2M^+(aq) + 2OH^-(aq) + H_2(g)$$

The alkali metals also react with nearly all the elemental nonmetals. Lithium, however, is the only one that reacts directly with N_2 to form an ionic nitride.

$$6Li(s) + N_2(g) \longrightarrow \underset{\text{lithium nitride}}{2Li_3N(s)}$$

Compounds with Oxygen Some of the most interesting reactions of the alkali metals are those that occur with molecular oxygen. All these metals react rapidly with O_2 when exposed to air, but the ion formed by oxygen varies for the different metals, as we studied earlier. Lithium reacts with oxygen to form the "normal oxide," Li_2O, which contains the O^{2-} ion.

The normal oxide of sodium can also be made, but the supply of oxygen must be limited. In an abundant supply of oxygen, sodium forms the yellowish white peroxide, Na_2O_2, that contains the peroxide ion, $O_2{}^{2-}$. (Sodium peroxide forms even when Na_2O is exposed to air.)

Peroxides and superoxides of the alkali metals are powerful oxidizing agents. Na_2O_2 is used as a bleach because it oxidizes many colored substances to give colorless products.

When the remaining alkali metals, potassium through cesium, are exposed to air, they form superoxides. These are yellow solids that contain the paramagnetic superoxide ion, $O_2{}^-$, and have the formula MO_2. An example is potassium superoxide, KO_2. The normal oxides of K, Rb, and Cs can also be made, but by indirect methods.

The reactions of the alkali metal oxides, peroxides, and superoxides with water are discussed in Chapter 1 of this supplement on page 16.

Chlorides, Hydroxides, and Carbonates Although the oxides of the alkali metals are interesting compounds, of much greater practical importance are the chlorides, hydroxides, and carbonates. And among these, the most important are those of sodium and potassium. This is simply because sodium and potassium are the most abundant of the Group IA metals.

As mentioned earlier, sodium chloride is a very common chemical. This makes NaCl an inexpensive raw material from which other compounds of sodium can be made. Potassium chloride is much less common. Its principal sources are the minerals *sylvite,* composed primarily of KCl, and *carnalite,* which has a composition corresponding to $KCl \cdot MgCl_2 \cdot 6H_2O$. These minerals are important fertilizers because plants require both potassium and magnesium ions.

Because sodium chloride is so plentiful, sodium compounds tend to be far less expensive to produce and therefore far more commercially important than similar compounds of the other alkali metals. A typical example is sodium hydroxide—industry's most important base. (NaOH usually ranks about eighth in total industrial production.) This compound, known as *lye* or *caustic soda* in commerce, is manufactured from NaCl by electrolysis, as described in Chapter 18. The overall reaction, you may recall, is

$$2NaCl(aq) + 2H_2O \xrightarrow{\text{electrolysis}} 2NaOH(aq) + Cl_2(g) + H_2(g)$$

Potassium hydroxide, KOH, can be made in the same way from KCl, but the amount produced annually is far less. One of its principal uses is as an electrolyte in batteries (for example, the alkaline Zn/MnO_2 battery and the alkaline Zn/HgO battery).

The carbonates represent another important group of compounds. For example, among sodium compounds, sodium carbonate (known as *soda ash* in commerce) usually ranks third behind NaOH in annual production. About half is used to make glass and the rest to make other chemicals, paper, detergents, and in water softening. Almost 80% of the Na_2CO_3 produced each year comes from a mineral called *trona,* $Na_2CO_3 \cdot NaHCO_3 \cdot 2H_2O$. The rest is manufactured from $NaHCO_3$, which is made from NaCl by the **Solvay process.**

The raw materials for the Solvay process are sodium chloride, ammonia, and calcium carbonate (limestone). The limestone is heated, causing it to decompose to CaO plus CO_2. The CO_2 from this reaction is bubbled into a cold solution of sodium chloride and ammonia. The dissolved CO_2 reacts with water and ammonia to form aqueous ammonium bicarbonate.

$$CO_2(aq) + H_2O + NH_3(aq) \longrightarrow NH_4^+(aq) + HCO_3^-(aq)$$

At this point, the solution contains the ions Na^+, Cl^-, NH_4^+, and HCO_3^-. At low temperatures, sodium bicarbonate is less soluble than any of the other possible salts that could be formed (NH_4Cl, NH_4HCO_3, or NaCl), so $NaHCO_3$ precipitates, leaving ammonium ion and chloride ion in the solution. The net overall reaction, therefore, can be represented by the molecular equation

$$NaCl(aq) + CO_2(aq) + H_2O + NH_3(aq) \longrightarrow NaHCO_3(s) + NH_4Cl(aq)$$

Ammonium chloride, which is recovered from the reaction mixture after the sodium bicarbonate is removed by filtration, is then allowed to react with the calcium oxide produced earlier by the decomposition of the limestone.

$$2NH_4Cl + CaO \longrightarrow CaCl_2 + 2NH_3 + H_2O$$

The ammonia formed in this reaction is recycled.

Most of the sodium bicarbonate from the Solvay process is converted to sodium carbonate by thermal decomposition according to the equation

$$2NaHCO_3(s) \xrightarrow{\text{heat}} Na_2CO_3(s) + H_2O + CO_2(g) \qquad (3.1)$$

Dry-chemical fire extinguishers often contain $NaHCO_3$ which decomposes by this reaction when sprayed on a fire. The products of the decomposition smother the flames.

Ordinarily, sodium carbonate is sold in two forms. One, called *washing soda,* is the hydrate $Na_2CO_3 \cdot 10H_2O$, which is formed when Na_2CO_3 is crystallized from water below 35.2 °C. The other, $Na_2CO_3 \cdot H_2O$, is formed when sodium carbonate is crystallized above this temperature.

Sodium bicarbonate, the primary product of the Solvay process, is itself an important chemical. Solutions of it are mildly basic and serve as a buffer because of the ability of the HCO_3^- ion to neutralize both acids and bases.

$$HCO_3^- + H_3O^+ \longrightarrow H_2CO_3 + H_2O$$

$$HCO_3^- + OH^- \longrightarrow CO_3^{2-} + H_2O$$

For this reason, $NaHCO_3$ is recommended as an additive for swimming pools where it helps control the pH of the water when other chemicals are added to destroy bacteria.

Sodium bicarbonate is commonly called *baking soda* or *bicarbonate of soda*. When added to dough that contains a mildly acidic ingredient, it produces a light and appealing product by reaction of the HCO_3^- ion with H^+ to give tiny bubbles of CO_2 that cause the dough to rise as it is baked.

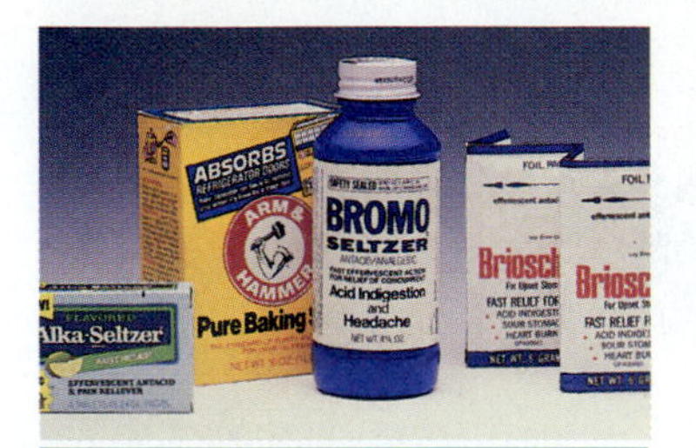

Some consumer products that contain sodium bicarbonate.

Sodium bicarbonate is also used in various analgesic products such as Alka Seltzer and Brioschi. The analgesic is usually aspirin; the $NaHCO_3$ provides the fizz and helps neutralize excess stomach acid.

Potassium carbonate, K_2CO_3, is known by the common name *potash* and is found in the ashes of burned plants and wood. It is used in making certain soft soaps, pottery, glass, and other compounds of potassium.

Solubilities of Salts

In the laboratory, sodium and potassium salts are common chemicals because nearly all are soluble in water. Therefore, if a particular anion is needed for a reaction in solution, a chemist almost always reaches for a bottle of its sodium or potassium salt. However, salts of these metals with a given anion are not equally soluble. An interesting and useful generalization is that the sodium salt of a *strong acid* is often more soluble than the potassium salt, whereas the potassium salt of a *weak acid* is often more soluble than the sodium salt. Table 3.2 contains some data that illustrate this.

Spectra

When ions of an alkali metal are added to a flame, brilliant colors are produced that are characteristic of the element's atomic spectrum. Sodium salts, for example, give a bright yellow flame (Figure 3.8). If you've ever softened a

TABLE 3.2 Solubilities of Some Sodium and Potassium Salts[a]

	Sodium Salt	Solubility (mol/100 g H_2O)	Potassium Salt	Solubility (mol/100 g H_2O)
Strong Acids				
Hydrochloric (HCl)	NaCl	0.61	KCl	0.46
Perchloric ($HClO_4$)	$NaClO_4$	1.49	$KClO_4$	0.01
Nitric (HNO_3)	$NaNO_3$	0.86	KNO_3	0.12
Weak Acids				
Acetic ($HC_2H_3O_2$)	$NaC_2H_3O_2$	1.45	$KC_2H_3O_2$	2.58
Tartaric ($H_2C_4H_4O_6$)	$Na_2C_4H_4O_6$	0.035	$K_2C_4H_4O_6$	0.64
Citric ($H_3C_6H_5O_7$)	$Na_3C_6H_5O_7$	0.25	$K_3C_6H_5O_7$	0.56

[a] Compared at the same temperature.

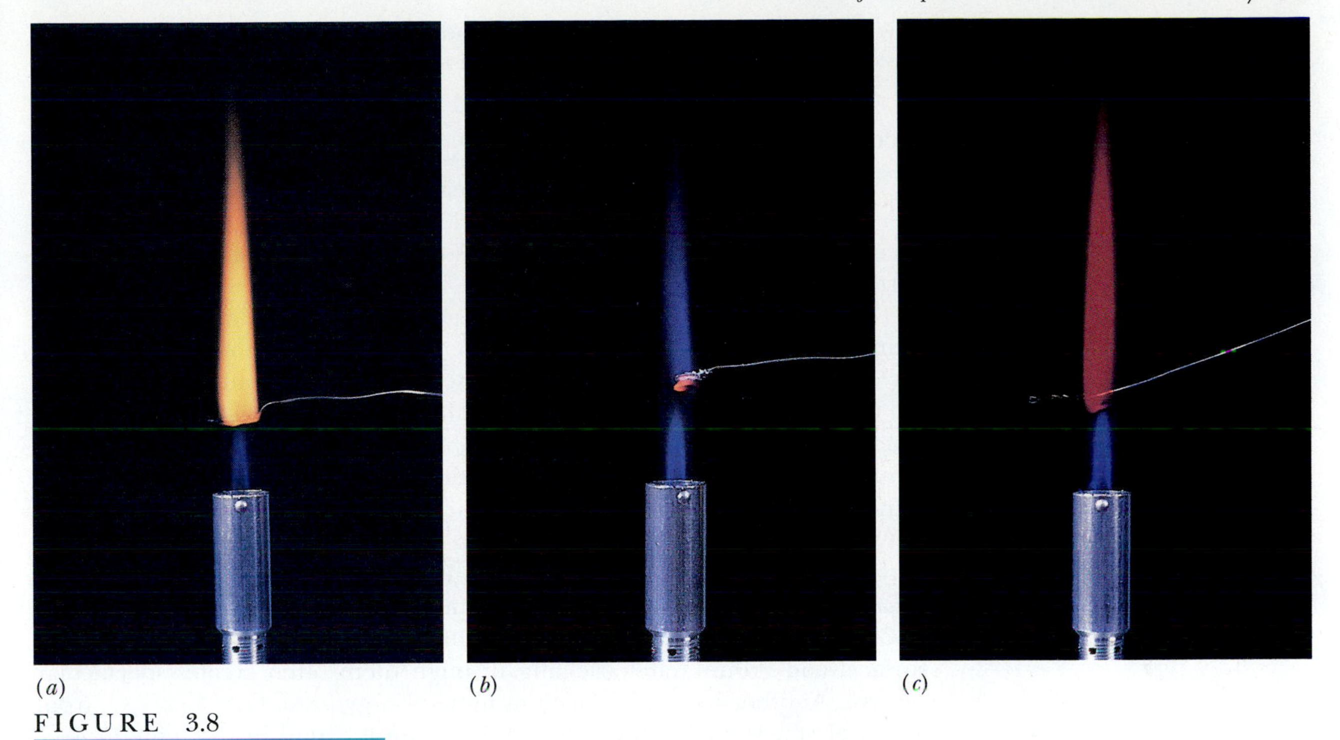
(*a*) (*b*) (*c*)

FIGURE 3.8
The colors imparted to a Bunsen burner flame by (*a*) sodium, (*b*) potassium, and (*c*) lithium.

glass rod or glass tubing in a Bunsen burner flame, you have seen this yellow color. It is produced by sodium ions that are vaporized from the hot glass.

Na_2CO_3 is used to make glass and is the source of the Na^+ ions that give the yellow color to a flame that's used to soften glass.

The yellow flame produced by sodium-containing compounds is so easily recognized that it forms the basis of a **flame test** for this element. If a sample is suspected to contain sodium, a clean wire is dipped into a solution of the sample and then held in a flame. A bright yellow color in the flame confirms the presence of sodium; the absence of the bright yellow flame means the sample contains no sodium.

The other alkali metals for which a flame test is normally used are lithium and potassium. Potassium salts impart a pale violet color to a flame, and lithium salts give a beautiful, deep red color. The pale violet of potassium is sometimes difficult to see, particularly if sodium is present. The yellow from the sodium is so intense that it masks the violet produced by the potassium. Viewing the flame through a special blue glass called *cobalt glass* filters out the yellow and allows the pale violet of the potassium flame to be seen.

Some fireplace logs are soaked in solutions of metal salts to produce brightly colored flames when they are burned.

3.4 THE METALS OF GROUP IIA: THE ALKALINE EARTH METALS

The elements of Group IIA are beryllium, magnesium, calcium, strontium, barium, and radium. Collectively they are known as the **alkaline earth metals** and, as with their neighbors in Group IA, there are strong similarities in chemical and physical properties among the members of the group. However, the similarities are not as striking as among the alkali metals. For example, lithium is much more like cesium than beryllium is like barium. This is the beginning of a general phenomenon that is observed as we move from either the left or right toward the representative elements in the center of the peri-

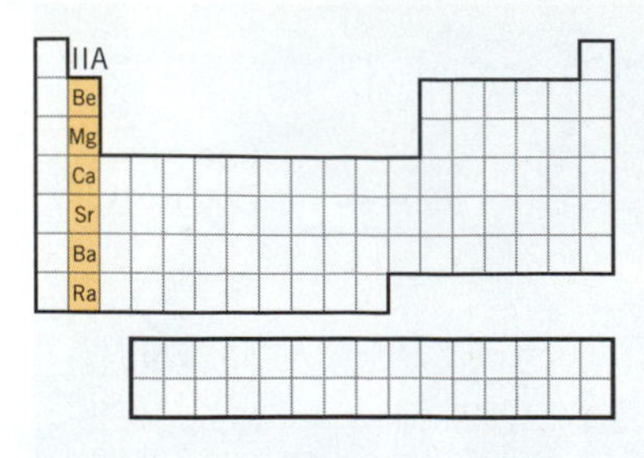

odic table; there is a progressively larger difference between the properties of the element at the top of the group and the element at the bottom.

Occurrence

The alkaline earth metals are almost as reactive as the metals of Group IA, and therefore they are never found in the elemental state in nature; they always occur in compounds, either in aqueous solution or in mineral deposits in the Earth.

Calcium is the third most abundant metal in the Earth's crust.

In seawater, the concentration of Mg^{2+} is 0.056 *M* and the concentration of Ca^{2+} is 0.011 *M*.

The most abundant of the alkaline earth metals are calcium and magnesium. As you learned earlier, one of the most common ions in seawater is Mg^{2+}. Calcium ions are also present in seawater, and marine organisms take the Ca^{2+} from the water to make their calcium carbonate shells. Calcium and magnesium ions are also found in large concentrations in certain underground brines. In the United States, particularly rich sources of these ions are brines that are pumped from wells in Michigan.

On land, the alkaline earth metals are found in various mineral deposits. Largest among them are limestone (calcium carbonate, $CaCO_3$) and *dolomite* (a mixed calcium magnesium carbonate, generally written as $CaCO_3 \cdot MgCO_3$). Many of these limestone deposits occur below the Earth's surface, and groundwater trickling through them often creates spectacular caverns. Another important calcium mineral is gypsum, $CaSO_4 \cdot 2H_2O$, from which plaster is made. The element beryllium is found in the mineral *beryl*, $Be_3Al_2(SiO_3)_6$. You may have seen beryl crystals such as *emerald* and *aquamarine* (Figure 3.9). Strontium and barium occur chiefly as their sulfates and carbonates, and radium is isolated from uranium ores because it is a product of the radioactive decay of uranium.

Radium was discovered by Marie and Pierre Curie.

Physical Properties

Some properties of the alkaline earth metals are given in Table 3.3. The atoms of these elements each lose two outer electrons to a metallic lattice and the doubly charged cations left behind are bound more tightly to the surrounding sea of electrons than are the singly charged cations in crystals of the alkali metals. As a result, the Group IIA metals have higher melting points and are harder than their neighbors in Group IA. They are also more dense than the Group IA metals because their atoms are smaller.

FIGURE 3.9
Naturally occurring emerald (*left*) and aquamarine (*right*). When cut and polished, they give beautiful gems. Both are forms of the mineral beryl, $Be_3Al_2(SiO_3)_6$.

TABLE 3.3 Some Properties of the Alkaline Earth Metals

Element	Electron Configuration	Ionization Energy (kJ/mol)		Density (g/cm^3)	Melting Point (°C)	Reduction Potential (V)[a]
		First	Second			
Beryllium	[He] $2s^2$	899	1757	1.85	1278	−1.85
Magnesium	[Ne] $3s^2$	737	1450	1.75	651	−2.37
Calcium	[Ar] $4s^2$	590	1145	1.55	843	−2.76
Strontium	[Kr] $5s^2$	549	1059	2.54	769	−2.89
Barium	[Xe] $6s^2$	503	960	3.51	725	−2.90
Radium	[Rn] $7s^2$	509	975	—	700	−2.82

[a] For the reaction: $M^{2+}(aq) + 2e^- \longrightarrow M(s)$.

The ionization energies of the alkaline earth metals are low, but higher than those of the alkali metals, so the Group IIA metals are not quite as easily oxidized as the alkali metals. This is reflected in the reduction potentials of the alkaline earth metals, which tend not to be quite as negative as the alkali metal alongside. Near the bottom of the group, however, the differences become negligible.

Preparation

The only alkaline earth metals to be produced in significant amounts are beryllium and magnesium. They are the only two that do not react readily with air and moisture at room temperature, so they are the only ones that can survive prolonged contact with our environment.

Beryllium is made by electrolysis of molten beryllium chloride. However, $BeCl_2$ is covalent and a poor conductor of electricity, so NaCl must be added to the melt to serve as an electrolyte. Metallic beryllium is used to make windows of X-ray tubes, because of all usable metals it is the most transparent to X rays. The metal is also used in alloys. One is a copper–beryllium alloy from which springs, electrical contacts, and tools are made. Hammers and wrenches made from Be/Cu alloys do not produce sparks when struck against steel and can therefore be used in explosive atmospheres. Beryllium compounds are quite toxic, and tough safety standards must be followed in fabricating beryllium-containing products.

Beryllium is used in alloys with copper and bronze to give them hardness.

Finely divided BeO is very toxic if inhaled.

Magnesium is recovered both from seawater and from ores obtained from the Earth—chiefly carnalite, $MgCl_2 \cdot KCl \cdot 6H_2O$, and dolomite, $CaCO_3 \cdot MgCO_3$. Recovery from dolomite involves first heating the mixed carbonate to give a mixed oxide.

Heating the carbonate in air like this is called *calcining*.

$$CaCO_3 \cdot MgCO_3 \xrightarrow{\text{heat}} CaO \cdot MgO + 2CO_2$$

This oxide mixture is added to seawater, which contains additional magnesium as Mg^{2+} ions. Hydrolysis of the oxides gives the hydroxides.

$$CaO + H_2O \longrightarrow Ca(OH)_2$$

$$MgO + H_2O \longrightarrow Mg(OH)_2$$

Calcium hydroxide is considerably more soluble than magnesium hydroxide, and the hydroxide ions from the $Ca(OH)_2$ cause all the Mg^{2+} ions in the seawater solution to be precipitated as $Mg(OH)_2$. The combined precipitate of $Mg(OH)_2$ is then allowed to settle in large ponds from which it is later

FIGURE 3.10

Metallic magnesium burns in air. The reaction produces a large amount of heat and light. The product of the reaction, magnesium oxide, can be seen dangling from the burning magnesium ribbon.

Magnesium hydroxide settling ponds at Dow Chemical Company's plant in Freeport, Texas, where magnesium is recovered from seawater.

recovered. As discussed earlier, the $Mg(OH)_2$ is dissolved in hydrochloric acid, and the resulting solution is evaporated to give solid $MgCl_2$, which is melted and electrolyzed to give metallic magnesium and chlorine.

Because magnesium has such a low density and a moderate strength, it is useful as a structural metal when alloyed with aluminum. Perhaps you have an aluminum–magnesium alloy ladder in your home. Another use of magnesium is in flashbulbs, fireworks, and incendiary bombs. Magnesium burns in air with the evolution of much heat and a very bright light, as shown in Figure 3.10. In a flashbulb, a thin magnesium wire is heated electrically by a battery and this ignites the metal, which burns very quickly in the pure oxygen atmosphere that surrounds it.

Barium is an important element in many of the high-temperature superconductors discovered in recent years.

Calcium, strontium, and barium have few commercial uses as metals other than as reducing agents in specialized metallurgical operations. They are produced in only limited quantities, usually by electrolysis of their molten chlorides.

Chemical Properties and Group Trends

As noted earlier, the alkaline earth metals are easily oxidized. In fact, all are oxidized easily enough to be attacked by water, at least in principle. However, the oxides of beryllium and magnesium have low solubilities in water. A surface coating of oxide therefore is able to protect these two metals from attack. The oxides of the elements below magnesium, however, are reasonably soluble in water. This permits water to come in contact with the metals, and they react to liberate hydrogen. For example, the reaction of calcium with water, shown in Figure 3.11, is

$$Ca(s) + 2H_2O \longrightarrow Ca(OH)_2(aq) + H_2(g)$$

FIGURE 3.11

A piece of calcium metal reacts with water to liberate H_2 gas, which rises to the surface as bubbles. The reaction also produces $Ca(OH)_2$.

The Group IIA metals do not react with water as violently as the alkali metals. Although magnesium is not attacked by cold water, it does react slowly with hot water and fairly rapidly with steam to form H_2 and $Mg(OH)_2$. Beryllium, on the other hand, does not react at all with water.

In their compounds, the alkaline earth metals always exist in the +2 oxidation state, corresponding to the "loss" of the outer pair of *s* valence electrons from each of their atoms. In general, compounds of calcium, strontium, and barium are distinctly ionic; that is, their compounds exhibit properties that are typical of ionic substances. Most magnesium compounds are also ionic, although magnesium does form some covalently bonded *organomagnesium* compounds in which the Mg atom is bonded to carbon atoms in groups derived from hydrocarbons. An example is $Mg(C_2H_5)_2$ in which two *ethyl groups,* C_2H_5, are covalently bonded through carbon to the magnesium atom. (Because Mg is less electronegative than C, it still is assigned an oxidation number of +2.)

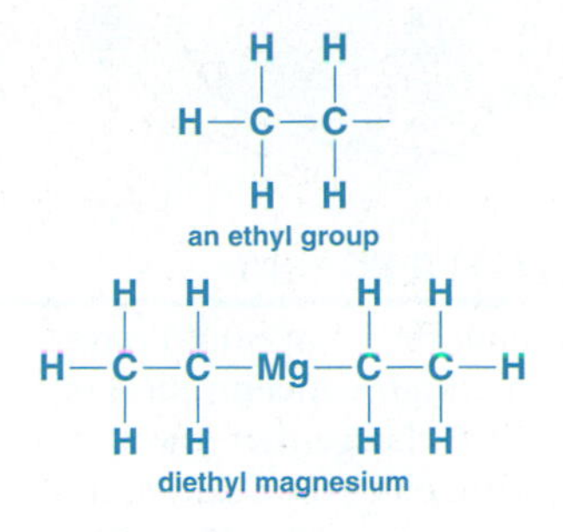

In the case of beryllium, there is no evidence that Be^{2+} ions actually exist. Beryllium compounds such as $BeCl_2$ are covalent. In fact, solid $BeCl_2$ really contains long strands of $BeCl_2$ units linked together by covalent bonds to give a structure that has the general formula $(BeCl_2)_x$, where *x* is a large number. The structure of these $BeCl_2$ chains is illustrated in Figure 3.12. We can view the structure as being formed by the donation of a pair of electrons from a chlorine of one $BeCl_2$ to the beryllium of another. These covalent bonds (we could call them *coordinate covalent bonds*) permit the Be atoms to complete their octets, and in the process, each Be atom becomes surrounded by four Cl atoms that are arranged approximately at the corners of a tetrahedron.

Recall that when one atom supplies both electrons in the formation of a covalent bond, the bond is called a coordinate covalent bond. Of course, once it is formed, the coordinate covalent bond is no different than any other covalent bond.

The large degree of covalent character in the bonds that beryllium atoms form with nonmetals is attributed to the small size and highly concentrated positive charge that a true Be^{2+} ion would have. As illustrated in Figure 3.13, if a Be^{2+} ion were placed next to an anion, its high concentration of positive charge would distort the anion's electron cloud and draw electron density into the region between the Be^{2+} and the anion. Because this would place the electron density *between* the two nuclei, it would produce, in effect, a covalent bond. Larger positive ions—Mg^{2+}, for example—are considerably less effective in distorting the electron cloud of an anion because the positive charge is more spread out. As a result, the bonds formed between Mg^{2+} and anions such as Cl^- are considerably less covalent, and therefore much more ionic.

As with the alkali metals, the alkaline earth metals combine directly with most of the nonmetals. Except for beryllium, the metals combine with nitrogen to form essentially ionic nitrides. For example,

$$3Mg(s) + N_2(g) \longrightarrow Mg_3N_2(s)$$

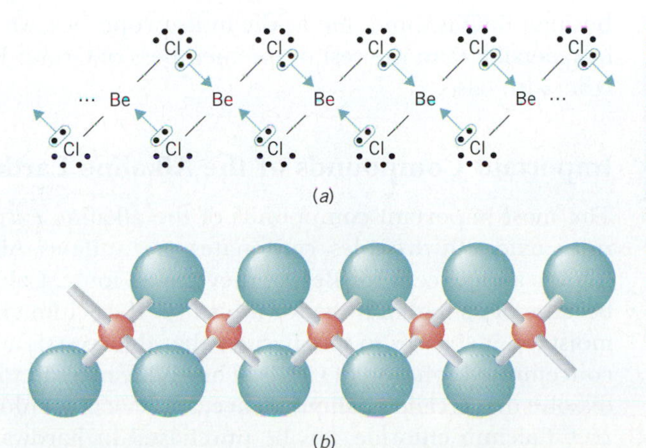

FIGURE 3.12

The formation of $(BeCl_2)_x$ from $BeCl_2$. (*a*) The formation of coordinate covalent bonds between the Cl atoms and Be atoms permits Be to complete its octet. The result is long chains of $BeCl_2$ units. (*b*) The arrangement of the chlorine atoms around each beryllium is approximately tetrahedral.

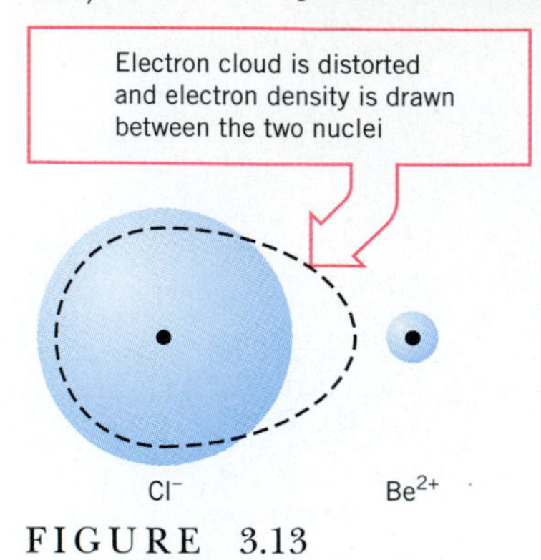

FIGURE 3.13

A small Be^{2+} ion would have such a high concentration of positive charge that it would distort the electron cloud of an anion such as Cl^-. The electron density drawn between the nuclei gives a great deal of covalent character to the bond.

Diagonal Similarities in the Periodic Table In some ways, lithium resembles magnesium more than it resembles the other members of Group IA. For example, lithium is alone among the alkali metals in its reaction with N_2. For lithium and magnesium, similarities include the formation of certain covalent compounds such as LiC_2H_5 and the $Mg(C_2H_5)_2$ mentioned above, as well as similarities in the solubilities of some of their salts. The parallel behavior of Li and Mg is often referred to as a *diagonal similarity* because of the relative locations of these elements in the periodic table, and it appears to be due to their similar ratios of ionic charge to ionic radius. This causes Li^+ and Mg^{2+} to have about the same effects on adjacent anions and therefore about the same degree of covalent character in their metal–nonmetal bonds. A similar diagonal similarity occurs for Be and Al, which we will discuss later.

The effect of small cation size and high cation charge on the covalent character and acid–base properties of metal oxides was discussed in Chapter 1 of this supplement.

Trends in Metallic Character Earlier we noted that as we descend a group in the periodic table, the elements become progressively more metallic. For example, in Group IVA the element at the top is a nonmetal (carbon) and the element at the bottom is a metal (lead). To a much lesser extent, we also observe this trend in Group IIA. All the alkaline earth metals are typically metallic in their physical properties; they are good conductors of heat and electricity, for example. But in some of their chemical properties, beryllium is considerably less metallic, and magnesium slightly less metallic than the heavier members of the group. For example, beryllium compounds are predominantly covalent—a property that we usually associate with the compounds formed by the nonmetallic elements. And we also find that magnesium forms some covalent compounds, too, but not nearly as many as beryllium.

Another example is the behavior of the metals and their oxides toward hydrogen ions. Earlier you learned that the oxides of metals are basic; either they dissolve in water to form basic solutions or they react with acids. On the other hand, typical oxides of nonmetals, such as SO_2, are acidic and either give acids in water or react with bases.

When we examine the behavior of the alkaline earth metals, we find that they all react with nonoxidizing acids such as HCl to liberate H_2, and their oxides react with HCl to give the corresponding salt and water. Beryllium and its oxide, however, are amphoteric and react with strong base as well.

$$Be(s) + 2H_2O + 2OH^-(aq) \longrightarrow Be(OH)_4^{2-}(aq) + H_2(g)$$

$$BeO(s) + H_2O + 2OH^-(aq) \longrightarrow Be(OH)_4^{2-}(aq)$$

Because BeO is somewhat acidic in its properties, we can say that beryllium is less metallic than the rest of the members of Group IIA, whose oxides do not react with base.

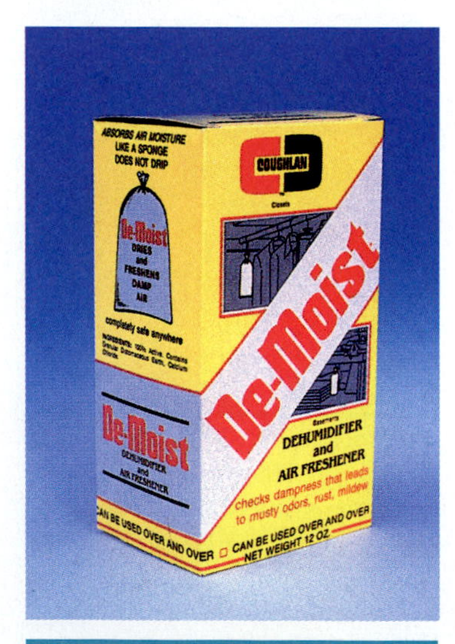

Calcium chloride has a strong affinity for water and readily absorbs moisture from the air. Products such as this, available in hardware stores, make use of this property to lower the humidity in damp basements or closets.

Important Compounds of the Alkaline Earth Metals

The most important compounds of the alkaline earth metals are their chlorides, oxides, hydroxides, carbonates, and sulfates. All the chlorides are water soluble and, except for $BeCl_2$, they are all ionic. Calcium chloride is unusual because of its high affinity for moisture. If calcium chloride is left exposed to moist air, it absorbs so much water that the crystals actually dissolve to form a concentrated solution of $CaCl_2$. The ability of a salt to absorb enough water to dissolve in it is called **deliquescence,** and calcium chloride is said to be *deliquescent.* Calcium chloride can be purchased in hardware stores (although not always under its chemical name) for use in removing moisture from places with high humidity, such as damp basements.

The oxides of the alkaline earth metals can be formed by direct combination of the elements.

$$2M(s) + O_2(g) \longrightarrow 2MO(s)$$

However, the usual method of preparation is by thermal decomposition of the carbonate. We have already seen this several times in the decomposition of limestone (calcium carbonate). Calcium oxide, known commercially as *lime* or sometimes *quick lime,* usually ranks about fourth in total tonnage produced annually. The reason is that it is inexpensive and a reasonably strong base. When treated with water, a process called *slaking,* it forms calcium hydroxide.

The *lime* that is spread on lawns or gardens to reduce soil acidity is really pulverized limestone or dolomite.

$$CaO(s) + H_2O \longrightarrow Ca(OH)_2(s)$$

$Ca(OH)_2$ is sometimes called *slaked lime.*

Calcium oxide is an important ingredient in portland cement, and this is one of the first reactions to occur when water is added to the cement.

Magnesium oxide can also be made from its carbonate by thermal decomposition. It is much less reactive toward water than CaO, especially if the solid is heated first to a high temperature. In fact, as noted earlier, magnesium metal is protected from attack by water by a thin film of MgO that forms on its surface. One of the principal uses of magnesium oxide is in special *refractory* bricks that withstand high temperatures in furnaces. Magnesium oxide is also used in making paper and in certain antacid preparations.

The hydroxides of the alkaline earth metals increase in solubility from $Mg(OH)_2$ to $Ba(OH)_2$.

MgO melts at about 2800 °C (approximately 5070 °F)!

Magnesium hydroxide is generally made by adding base to a solution containing Mg^{2+}. This precipitates $Mg(OH)_2$.

$$Mg^{2+}(aq) + 2OH^-(aq) \longrightarrow Mg(OH)_2(s)$$

The antacid and laxative known as *milk of magnesia* is a suspension of $Mg(OH)_2$ in water.

The most important sulfates of the alkaline earth metals are those of magnesium, calcium, and barium. The solubilities of the sulfates decrease going down the group; $MgSO_4$ is quite soluble, whereas $BaSO_4$ is very insoluble. Magnesium sulfate as its hydrate $MgSO_4 \cdot 7H_2O$ is known as *Epsom salts,* and is used in the tanning of leather, to treat fabrics so they readily accept dyes, to fireproof fabrics, as a fertilizer, and medicinally.

$BaSO_4$ $K_{sp} = 1.1 \times 10^{-10}$
$CaSO_4$ $K_{sp} = 2.4 \times 10^{-5}$

The dihydrate of calcium sulfate, $CaSO_4 \cdot 2H_2O$, is called *gypsum.* Partial dehydration of gypsum gives plaster of Paris.

$$\underset{\text{gypsum}}{2CaSO_4 \cdot 2H_2O(s)} \xrightarrow{\text{heat}} \underset{\text{plaster of Paris}}{(CaSO_4)_2 \cdot H_2O(s)} + 3H_2O(g)$$

Usually, the formula for plaster of Paris is written as $CaSO_4 \cdot \frac{1}{2}H_2O$, which expresses the same mole ratio of $CaSO_4$ to H_2O as $(CaSO_4)_2 \cdot H_2O$.

One of the important uses of barium sulfate is in obtaining medical X-ray photographs of the digestive tract. A patient drinks a suspension of $BaSO_4$ in water and then an X-ray photograph is taken. The path of the patient's digestive tract is clearly visible on the film because $BaSO_4$ is opaque to X rays (see Figure 3.14). Even though the barium ion, like most heavy metal ions, is very toxic to the human body, barium sulfate is safe to drink because its solubility is so low. Hardly any Ba^{2+} is absorbed by the body as the $BaSO_4$ passes through the digestive system. Other uses of $BaSO_4$ are based on its whiteness; it is used as a whitener in photographic papers and as a filler in papers and polymeric fibers.

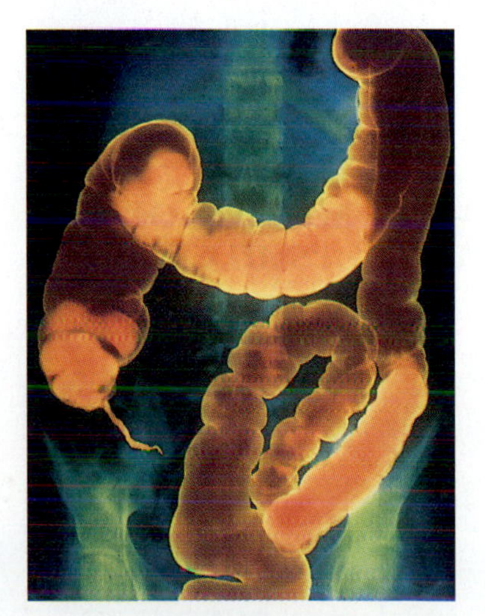

FIGURE 3.14
Barium sulfate, which is opaque to X rays, defines the path of the large intestine in a patient who has been given a "barium enema."

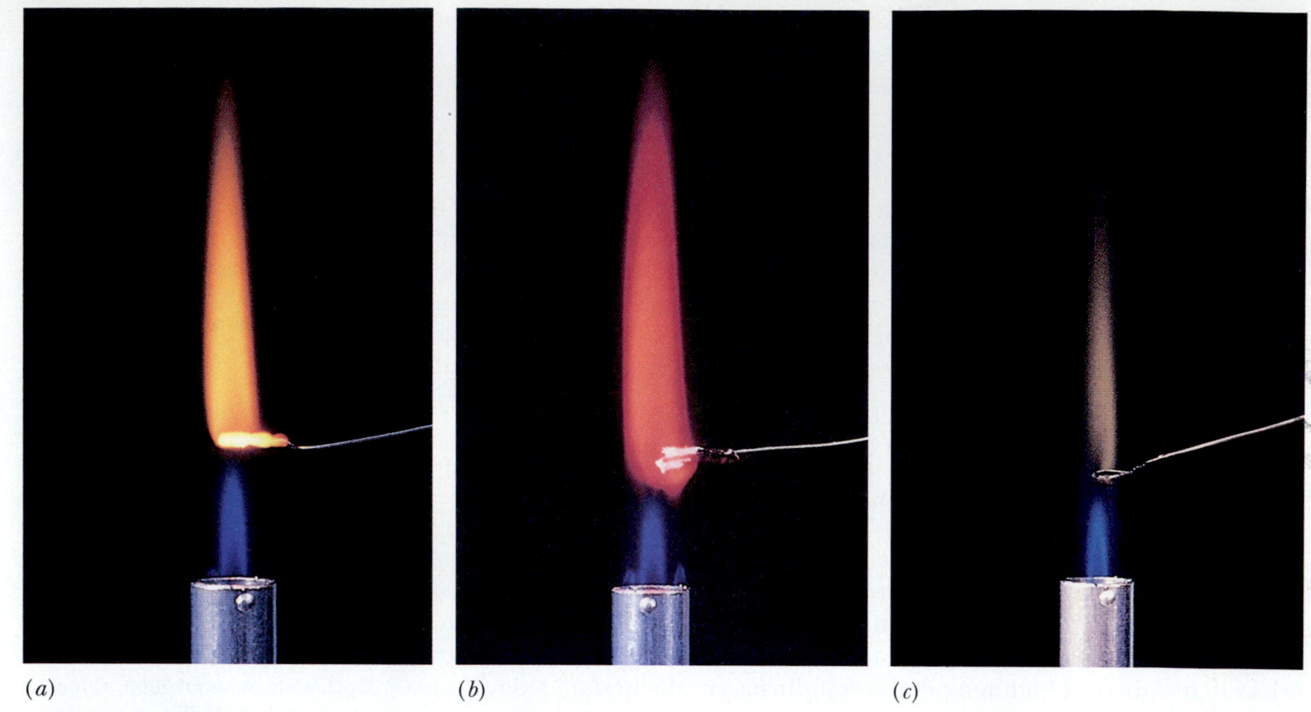

FIGURE 3.15

The colors given to a Bunsen burner flame by (*a*) calcium, (*b*) strontium, and (*c*) barium.

Spectra

Like the alkali metals, certain of the alkaline earth metals give characteristic colors to flames. Calcium salts give an orange-red color, strontium salts produce a bright red (crimson) flame, and barium salts give a yellow-green color. These are shown in Figure 3.15, and are intense enough to serve as flame tests. Like salts of the alkali metals, salts of these alkaline earth metals are used in coloring fireworks displays.

3.5 THE METALS OF GROUPS IIIA, IVA, AND VA

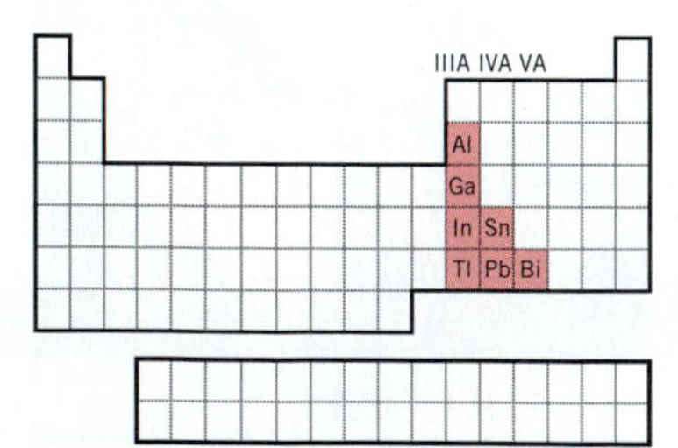

Except for aluminum, the metals in Groups IIIA, IVA, and VA are called *post-transition metals* because they follow the row of transition elements in their respective periods. The post-transition metals are, in general, considerably less reactive than the metals in Groups IA and IIA. They each have a completed *d* subshell just below the valence shell. Because these *d* electrons are less than 100% effective in screening the nuclear charge, the outer electrons of these elements are held much more tightly than the outer electrons of the alkali or alkaline earth metals.

The Metals of Group IIIA

The elements of Group IIIA are boron, aluminum, gallium, indium, and thallium. They have no general group name as do the elements of Groups IA and IIA. Except for boron, which is a metalloid, all are metals. Some of their properties are given in Table 3.4. Notice that gallium has a melting point of only 29.8 °C; it will melt in the palm of your hand (Figure 3.16). Gallium also has a very high boiling point, and this large liquid range has made gallium useful for certain types of thermometers.

TABLE 3.4 Some Properties of the Group IIIA Elements

Element	Electron Configuration	Ionization Energy (kJ/mol)			Density (g/cm^3)	Melting Point (°C)	Reduction Potential (V)[a]
		First	Second	Third			
Boron	[He] $2s^2 2p^1$	801	2427	3660	2.34	2200	—
Aluminum	[Ne] $3s^2 3p^1$	577	1816	2744	2.70	600	−1.66
Gallium	[Ar] $3d^{10} 4s^2 4p^1$	578	1971	2950	5.90	29.8	−0.53
Indium	[Kr] $4d^{10} 5s^2 5p^1$	559	1813	2690	7.31	157	−0.33
Thallium	[Xe] $4f^{14} 5d^{10} 6s^2 6p^1$	589	1961	2860	11.85	303	+0.72

[a] For $M^{3+}(aq) + 3e^- \rightarrow M(s)$.

Aluminum The only really important metal in Group IIIA, as far as we are concerned, is aluminum. On a weight basis, it is the third most abundant *element* in the Earth's crust, exceeded only by oxygen and silicon. This also means that it is the most abundant *metal* in the Earth's crust. It is often found combined with silicon and oxygen in *aluminosilicates,* which occur in various kinds of rock, such as granite, and in clays. Unfortunately, no one has found an economical way of separating the aluminum from these minerals, so they are not considered aluminum ores. Aluminum is also found in the mineral *cryolite,* which has the formula Na_3AlF_6 and contains the complex ion AlF_6^{3-}. The chief ore of aluminum is *bauxite,* which contains aluminum in the form of a hydrated oxide, $Al_2O_3 \cdot xH_2O$. The purification of bauxite is described on page 83, and the production of aluminum by the Hall–Héroult process is discussed in Chapter 18. As you know, aluminum has many uses, from aluminum foil to airplane construction. Many of its structural uses are based on its low density ("light weight") and moderate strength.

The Earth's crust consists of about 8.8% aluminum by weight.

In its compounds, aluminum always exists in the +3 oxidation state, which corresponds to the "loss" of its three valence electrons. However, an Al^{3+} ion is small and highly charged, and as we saw in the case of beryllium, this can cause considerable distortion of the electron cloud of a neighboring anion and impart a large degree of covalent character to the metal–nonmetal bond. For this reason, many of the compounds of aluminum show substantial covalent character in much the same way as compounds of beryllium.

Aluminum metal is easily oxidized; the reduction potential of $Al^{3+}(aq)$ equals −1.66 V. Fortunately, the reaction between aluminum and oxygen produces a tough oxide coating that adheres tightly to the aluminum surface and

FIGURE 3.16

The melting point of gallium is only 29.8 °C. Body temperature is 37 °C, which is high enough to cause the metal to melt in the palm of your hand.

SPECIAL TOPIC 3.1 / THE REACTIVITY OF ALUMINUM: A PROBLEM IN DENTISTRY

In capping a tooth, dentists often use a temporary aluminum cover to protect the tooth while the dental laboratory fabricates the permanent crown. If the tooth being treated has also received a fresh silver filling, however, the aluminum temporary crown can't be employed.

A silver filling is a mixture of silver powder and mercury. Mercury dissolves the silver to form an *amalgam*—a solution of a metal in mercury. This cements the silver particles together. When the silver filling is fresh, however, the silver particles still have mercury on their surfaces. It hasn't yet been absorbed by the silver. If an aluminum crown is placed over the tooth, the mercury can contact the aluminum and form an amalgam with it. If this happens, it prevents the adhesion of the protective Al_2O_3 coating and allows aluminum metal in the amalgam to react with oxygen and water in the patient's mouth. The resulting exothermic reaction can make the aluminum crown painfully hot.

protects the metal beneath from further attack. In fact, *anodized aluminum* has an oxide coating that is made deliberately thick by electrolysis. Because of the way it is formed, this coating is porous enough to accept and hold printing inks that would not otherwise stick to the aluminum.

The strong affinity of aluminum for oxygen can be shown experimentally by first dipping the aluminum into acid, which dissolves the oxide coating, and then into a solution of $HgCl_2$. In this second solution, some aluminum dissolves and metallic mercury deposits.

$$3Hg^{2+}(aq) + 2Al(s) \longrightarrow 3Hg(l) + 2Al^{3+}(aq)$$

The liquid mercury forms a solution (called an amalgam) with the aluminum at the surface. Aluminum oxide doesn't adhere to the amalgam, and this allows aluminum and oxygen to continue to react—a reaction that generates a great deal of heat and rapidly warms the aluminum to the point where it is too hot to hold (see Special Topic 3.1).

Aluminum oxide is an extremely stable compound and has a very exothermic heat of formation ($\Delta H_f^\circ = -1670$ kJ/mol). This fact accounts for some interesting uses of the metal. For example, one of the most spectacular reactions of aluminum is called the **thermite reaction,** in which iron oxide is reduced to metallic iron by aluminum.

$$2Al(s) + Fe_2O_3(s) \longrightarrow Al_2O_3(l) + 2Fe(l)$$

FIGURE 3.17
The thermite reaction is being used here to produce molten iron that flows into a mold between rail ends, thereby welding the rails together.

The thermite reaction has $\Delta H^\circ = -847.6$ kJ.

The large exothermic heat of formation of Al_2O_3 causes this reaction to be very exothermic, and enough heat is released to raise the temperature to nearly 3000 °C, a temperature at which the products are molten. The hot molten iron formed in the thermite reaction is often used to weld iron and steel parts together, as we see in Figure 3.17.

Another relatively recent use of the large amount of energy released by the formation of aluminum oxide is in the solid booster rockets used in launching the space shuttle, as seen in Figure 3.18. These boosters stand about 125 feet high and are 12 feet in diameter. The solid propellant in them is prepared by mixing powdered aluminum (the fuel), ammonium perchlorate (NH_4ClO_4, the oxidizer), a small amount of iron oxide powder (a catalyst), and a plastic epoxy resin binder. When first mixed, the propellant has the consistency of peanut butter and a gray color. It is poured into large casting molds and then cured slowly over a period of about 4 days to give a brick-red material with the consistency of a hard rubber eraser. When this rubbery propellant is ignited, it burns fiercely as the ammonium perchlorate oxidizes the aluminum to Al_2O_3.

FIGURE 3.18

Clouds of white aluminum oxide form in the exhaust from the booster rockets that help lift the space shuttle *Discovery* from its launch pad at Cape Canaveral, Florida.

FIGURE 3.19

In this closeup photo of the drain cleaner Drano, we see tiny bits of aluminum metal surrounded by crystals of sodium hydroxide that are colored slightly blue with a dye. In action, the aluminum bits react with the dissolved NaOH, forming bubbles of H_2 that cause a stirring effect in the clogged drain.

The large amount of energy that's released heats the gases formed by the reduction of the NH_4ClO_4 to very high temperatures. These hot gases expand with great force and help lift the space shuttle from its launch pad.

Aluminum is amphoteric. It dissolves in both acids and bases with the liberation of hydrogen. The equations are often written as follows.

$$2Al(s) + 6H^+(aq) \longrightarrow 2Al^{3+}(aq) + 3H_2(g) \quad \text{(acidic solution)}$$

$$2Al(s) + 2OH^-(aq) + 2H_2O \longrightarrow 2AlO_2^-(aq) + 3H_2(g) \quad \text{(basic solution)}$$

The second reaction illustrates why oven cleaners that contain lye (NaOH) should not be used on aluminum pots and pans—the sodium hydroxide attacks the aluminum and dissolves it. This reaction is exploited, however, by the drain cleaner Drano, which consists mostly of NaOH along with small bits of aluminum metal (Figure 3.19). When Drano is put into a clogged drain, the bubbles caused by the release of hydrogen stir the mixture and help it dissolve grease or hair that might be causing the stoppage.

The nature of the aluminum species in an aqueous solution is more complex than indicated by the equations above. For example, in acidic or neutral solutions, the aluminum ion exists as the complex ion $Al(H_2O)_6^{3+}$, in which six water molecules are bound rather tightly to the Al^{3+} ion in an octahedral arrangement. When aluminum salts crystallize from water, they usually contain this complex ion within their crystals.

Aqueous solutions that contain aluminum salts such as $AlCl_3$ or $Al_2(SO_4)_3$ are slightly acidic because the $Al(H_2O)_6^{3+}$ ion reacts with water. The attraction of the Al^{3+} ion for the electrons of the water molecules that surround it in the $Al(H_2O)_6^{3+}$ ion pulls some of the electron density away from the O—H bonds, as illustrated in Figure 3.20. This tends to further polarize the already polar O—H bonds of the water molecules, which increases the amount of positive charge on the hydrogen atoms. In turn, this makes it easier to remove the hydrogens as H^+ ions, and these become attached to other water mole-

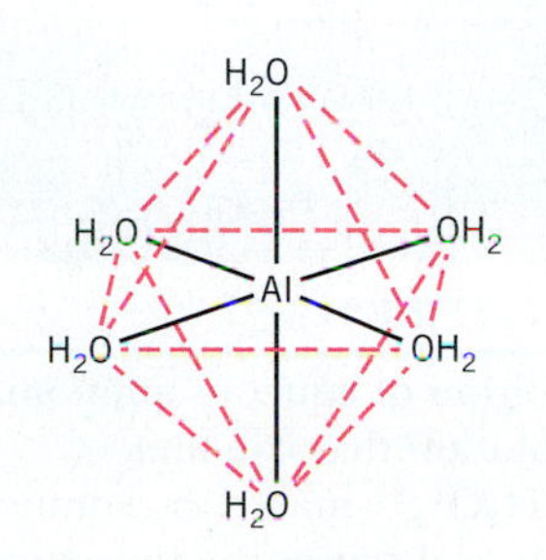

The $Al(H_2O)_6^{3+}$ ion.

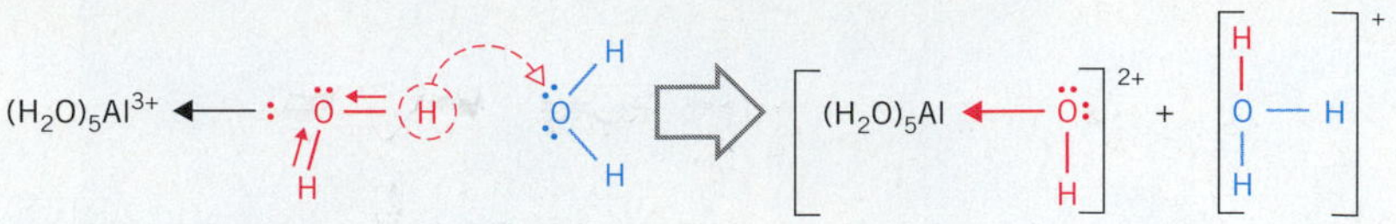

FIGURE 3.20

The aluminum ion draws electrons to itself from the oxygen atoms of the neighboring water molecules, which further polarizes the O—H bonds. This polarization, shown here for one of the six water molecules, makes it easier for an H^+ to be transferred to a molecule of water in the surrounding solvent, thereby producing H_3O^+ in the solution.

cules in the solvent to give H_3O^+ ions. The reaction is an equilibrium that can be represented by the equation

$$Al(H_2O)_6^{3+}(aq) + H_2O \rightleftharpoons Al(H_2O)_5(OH)^{2+}(aq) + H_3O^+(aq)$$

The addition of a base to a solution of an aluminum salt precipitates a *gelatinous* (gelatinlike) aluminum hydroxide (Figure 3.21). This can be formulated as a neutralization reaction in which hydroxide ions strip protons from three of the water molecules that are attached to the aluminum ion, turning them into hydroxide ions.

$$Al(H_2O)_6^{3+}(aq) + 3OH^-(aq) \longrightarrow Al(H_2O)_3(OH)_3(s) + 3H_2O$$

Aluminum hydroxide is amphoteric because it dissolves in either acid or base. When it dissolves in base, a hydroxide ion removes still another proton from a water molecule of the $Al(H_2O)_3(OH)_3$, and when it dissolves in acid, the hydronium ion adds a proton to an OH^- that's attached to the Al^{3+}. This continues until the $Al(H_2O)_6^{3+}$ ion is formed.

The hydroxide ions attached to the aluminum in $Al(H_2O)_3(OH)_3$ originally were water molecules in the $Al(H_2O)_6^{3+}$ ion.

Reaction with Base

$$Al(H_2O)_3(OH)_3(s) + OH^-(aq) \longrightarrow Al(H_2O)_2(OH)_4^-(aq) + H_2O$$

Reaction with Acid

$$Al(H_2O)_3(OH)_3(s) + H_3O^+(aq) \longrightarrow Al(H_2O)_4(OH)_2^+(aq) + H_2O$$

$$Al(H_2O)_4(OH)_2^+(aq) + H_3O^+(aq) \longrightarrow Al(H_2O)_5(OH)^{2+}(aq) + H_2O$$

$$Al(H_2O)_5(OH)^{2+}(aq) + H_3O^+(aq) \longrightarrow Al(H_2O)_6^{3+}(aq) + H_2O$$

Notice that the ion formed in basic solution, $Al(H_2O)_2(OH)_4^-$, is equivalent in terms of total atoms and charge to $AlO_2^- + 4H_2O$. Writing the formula AlO_2^- for the aluminum-containing species in basic solution, as we did earlier, is really shorthand for the more complicated substances that actually exist in the solution.

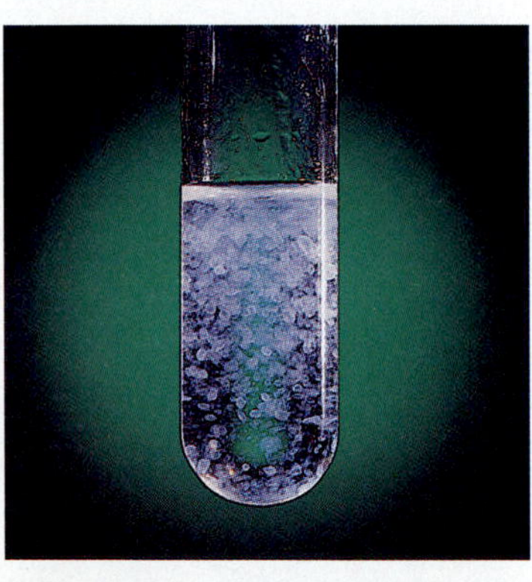

FIGURE 3.21

Addition of aqueous ammonia to a solution that contains $Al(H_2O)_6^{3+}$ makes the solution basic and causes the formation of a gelatinous precipitate of $Al(H_2O)_3(OH)_3$. The solid is usually called aluminum hydroxide and is often written without the water as $Al(OH)_3$.

Compounds of Aluminum Among the important compounds of aluminum are the oxide, the halides, and the sulfates. Aluminum oxide has the formula Al_2O_3, but it occurs in two different crystalline forms that differ greatly in their chemical reactivities. If the gelatinous hydroxide is heated mildly, it loses water and gives an oxide referred to as γ-Al_2O_3.

$$2Al(H_2O)_3(OH)_3 \xrightarrow{\text{heat}} \gamma\text{-}Al_2O_3 + 9H_2O$$

This form of the oxide readily dissolves in both acidic and basic solutions. If γ-Al_2O_3 is heated to temperatures above 1000 °C, its crystal structure changes

FIGURE 3.22

Natural gemstones of ruby (*left*) and sapphire (*right*). Both are composed of nearly pure aluminum oxide, Al_2O_3. The ruby is Al_2O_3 that contains traces of chromium. In the sapphire the Al_2O_3 is contaminated by traces of iron and titanium, which give it its color.

to that of α-Al_2O_3. This form of the oxide is very resistant to chemical attack. Naturally occurring α-Al_2O_3 is called *corundum*. Its crystals are very hard and it is commonly used as an abrasive in sandpaper. Aluminum oxide has a very high melting point (2045 °C), and it is also used to make refractory bricks for the interiors of furnaces.

Large crystals of corundum that contain traces of certain other metals are valued as gems. For example, sapphire is Al_2O_3 that contains small amounts of iron and titanium; ruby is Al_2O_3 containing trace amounts of Cr^{3+} (Figure 3.22). Artificial sapphires and rubies are made by melting very pure aluminum oxide with carefully measured amounts of the appropriate metal oxides and then cooling the mixture in such a way that large crystals are formed. Such artificial gems are difficult to distinguish from the naturally occurring variety, and they find uses as bearings in mechanical devices as well as in jewelry.

A compound of aluminum produced in very large quantities, much of which ultimately is converted to the hydroxide, is aluminum sulfate. It is usually made by dissolving bauxite in sulfuric acid.

$$Al_2O_3(s) + 3H_2SO_4(aq) \longrightarrow Al_2(SO_4)_3(aq) + 3H_2O$$

When aluminum sulfate is crystallized, it forms hydrates with as many as 18 water molecules per formula unit, $Al_2(SO_4)_3 \cdot 18H_2O$.

A major consumer of aluminum sulfate is the paper industry, where it is used to adjust acidity [recall that solutions of $Al_2(SO_4)_3$ are acidic] and to make the paper water resistant. Aluminum sulfate is also used in municipal water treatment, where it is added to the water along with lime (CaO). The calcium oxide reacts with water to make the solution slightly basic, and this causes gelatinous aluminum hydroxide, $Al(H_2O)_3(OH)_3$, to precipitate. As the precipitate settles to the bottom of the treatment tank, it carries with it finely suspended solids and bacteria.

When a solution that contains equal numbers of moles of aluminum sulfate and sodium sulfate is evaporated, crystals are formed that have the composition $NaAl(SO_4)_2 \cdot 12H_2O$. Similar crystals are formed from solutions that contain both aluminum sulfate and either ammonium sulfate or potassium sulfate. Their formulas are $NH_4Al(SO_4)_2 \cdot 12H_2O$ and $KAl(SO_4)_2 \cdot 12H_2O$. These are examples of **double salts,** which are compounds that contain the components of *two* salts in a definite ratio. A double salt that has the general formula $M^+M^{3+}(SO_4)_2 \cdot 12H_2O$ is called an **alum** and salts of this type include double salts formed not only by aluminum, but also by Cr^{3+} and Fe^{3+}. Under carefully

$M^+ = Na^+, K^+, NH_4^+$
$M^{3+} = Al^{3+}, Cr^{3+}, Fe^{3+}$

controlled conditions, large well-formed crystals of the alums can be grown that are quite beautiful (Figure 3.23).

FIGURE 3.23

A crystal of potassium alum, $KAl(SO_4)_2 \cdot 12H_2O$, shown in actual size. In the general formula for an alum, M^+ can be Na^+, K^+, or NH_4^+; M^{3+} can be Al^{3+}, Cr^{3+}, or Fe^{3+}.

Potassium alum, $KAl(SO_4)_2 \cdot 12H_2O$, is used to treat cotton fibers to make them absorb dyes more easily. Sodium alum, $NaAl(SO_4)_2 \cdot 12H_2O$, is another compound that is used along with sodium bicarbonate to make baking powders. When alum is added to moist dough, the acidity of the aluminum ion in water causes carbon dioxide to be released by reaction of hydronium ion with the bicarbonate ion.

$$H_3O^+(aq) + HCO_3^-(aq) \longrightarrow 2H_2O + CO_2(g)$$

The small bubbles of CO_2 cause the dough to rise and give the finished baked product a light and airy texture.

The anhydrous halides of aluminum are interesting because of similarities with the halides of beryllium; there is a substantial degree of covalent bonding between aluminum and the halide in the solid and particularly in the vapor. For example, aluminum bromide exists in the solid and vapor as Al_2Br_6—a *dimer* formed from two $AlBr_3$ units. Similarly, in the vapor aluminum chloride exists in the molecular form Al_2Cl_6. The structures of these molecules are illustrated in Figure 3.24. Their formation from AlX_3 (X = Cl or Br) can be viewed as an attempt by aluminum to complete its octet.

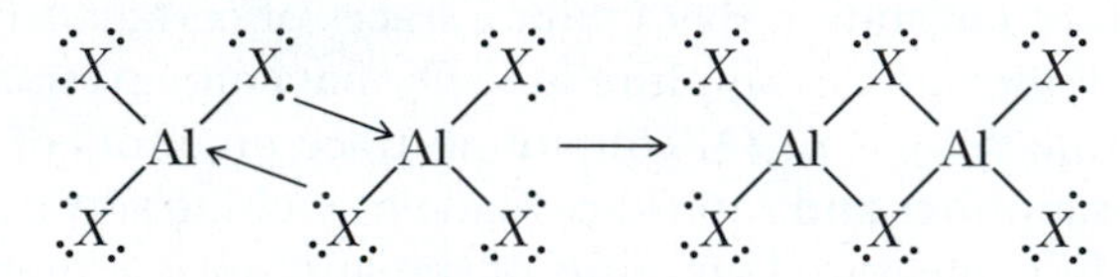

In the dimer, aluminum is surrounded by four pairs of electrons, and the halogens are arranged approximately at the corners of a tetrahedron with the aluminum at the center. This is, of course, what is expected based on the VSEPR theory. Note the similarity between the structure of the dimeric Al_2X_6 and the structure of the polymeric $(BeCl_2)_x$ illustrated in Figure 3.12.

When anhydrous aluminum bromide or chloride is dissolved in water, their molecules dissociate and the hydrated aluminum ion, $Al(H_2O)_6^{3+}$, is formed along with the corresponding halide ions. These reactions are quite exothermic.

Gallium, Indium, and Thallium These elements each have an outer electron configuration that can be represented as ns^2np^1. Each forms compounds in both the +1 and +3 oxidation states. The lower one corresponds to the loss of the outer p electron, and the higher one to the loss of the pair of s electrons as well. The most noteworthy aspect of the chemistry of these metals is the increasing tendency, going down the group, to form an ion with a 1+ charge. The reason appears to be that as the atoms become larger, the additional energy needed to remove the pair of s electrons is more than can be easily recovered by bond formation, so as the atoms become larger going down the group there is a lesser tendency to form the higher oxidation state, and an increasing tendency to form the lower one. Thus, for Ga the +3 state is common and the +1 state is rare, whereas for Tl both the +1 and +3 states are well known.

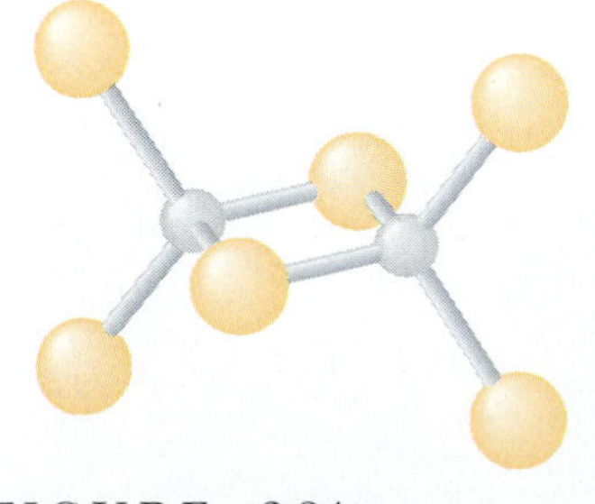

FIGURE 3.24

The molecular structure of Al_2X_6, where X can be Cl or Br.

The Metals of Groups IVA and VA

The metals of Group IVA are tin and lead. Bismuth is the most metallic element in Group V. Some of the properties of these elements are given in Table 3.5.

TABLE 3.5 Some Properties of Tin, Lead, and Bismuth

Element	Electron Configuration	Density (g/cm^3)	Melting Point (°C)	Oxidation States (Most Stable in Bold Type)
Tin	[Kr] $4d^{10}5s^2\,5p^2$	5.75	232	+2, **+4**
Lead	[Xe] $4f^{14}5d^{10}6s^26p^2$	11.35	328	**+2**, +4
Bismuth	[Xe] $4f^{14}5d^{10}6s^26p^3$	9.75	271	**+3**, +5

Tin is usually found in nature as its oxide, SnO_2, and can be reduced to the free metal by reaction with carbon (coke).

$$SnO_2 + C \xrightarrow{\text{heat}} Sn + CO_2$$

Often it is purified further electrolytically in the same manner as copper. The impure tin is made to be the anode of an electrolytic cell, and pure tin is made to be the cathode. As the cell operates, the impure anode gradually dissolves and pure tin is deposited on the cathode.

Elemental tin occurs in three allotropic forms. The most common is called *white tin* and is the shiny tin coating you find over steel on a "tin can." If tin is kept for long periods below a temperature of 13.2 °C, the white tin gradually changes to a powdery, nonmetallic form called *gray tin.* Tin objects kept in cold climates for long periods sometimes develop lumps as this change begins to take place, and at one time it was thought that some organism was attacking the tin. The phenomenon was called "tin disease." A third allotrope called *brittle tin* is obtained when white tin is heated; its properties are obvious from its name.

Lead occurs as a sulfide, PbS, in the ore *galena.* Roasting in air gives the oxide, which is then reduced by carbon to give the free metal.

$$2PbS + 3O_2 \xrightarrow{\text{heat}} 2PbO + 2SO_2$$

$$2PbO + C \xrightarrow{\text{heat}} 2Pb + CO_2$$

The lead from this process often contains silver, gold, and other metals as impurities, and it too can be purified by electrolysis. The silver and gold are recovered to help offset the cost of the electricity.

The Latin name for lead is *plumbum,* and the terms *plumbing* and *plumber* come from the early use of lead for pipes and pipe joints. The metal is also used to make batteries and solder and to manufacture tetraethyllead, $Pb(C_2H_5)_4$, a gasoline octane booster. The use of lead-containing additives in gasoline has been phased out in the United States because of the environmental dangers. However, tetraethyllead is still used in other countries. Special Topic 3.2 describes some of the problems of lead poisoning.

Bismuth is sometimes found in nature as the free metal, but usually it occurs as the oxide or sulfide. Roasting followed by reduction with carbon gives free bismuth.

$$2Bi_2S_3 + 9O_2 \xrightarrow{\text{heat}} 2Bi_2O_3 + 6SO_2$$

$$2Bi_2O_3 + 3C \xrightarrow{\text{heat}} 4Bi + 3CO_2$$

It is also a by-product of the production of lead because bismuth is often an impurity in lead ores.

SPECIAL TOPIC 3.2 / LEAD POLLUTION

Tetraethyllead, $Pb(C_2H_5)_4$, and similar lead compounds were long used to improve the performance of gasoline in vehicle engines. Although only about 3 grams of this additive were present per gallon of "leaded" gasoline, over a long enough time lead deposits would build up in engine cylinders. To prevent this, another additive—ethylene bromide—was used.

In scavenging lead from burning gasoline, ethylene bromide converted the lead into compounds volatile enough to be carried out in the exhaust. Of course, lead compounds then entered the atmosphere, from which they soon precipitated. And that's the rub. Lead compounds are poisons whether inhaled on dust or taken in food or drink. In the early 1970s, street sweepings in New York City contained over 2500 ppm of lead (as compared with the average abundance in the Earth's crust of 16 ppm). Rainfall along the southern California coast was once found to have 40 μg Pb/L. (The "grounds-for-rejection" limit on drinking water is 15 μg Pb/L.)

Alarmed government health officials working with elected people brought about a law requiring new cars to burn unleaded gasoline only. Such cars also are equipped with catalytic mufflers to reduce emissions of carbon monoxide, unburned hydrocarbons, and nitrogen monoxide; and lead compounds poison the catalyst in these mufflers. So it was required that gas tanks be made that would not be able to accept the larger gas nozzles for leaded gasoline. Until relatively recently, leaded gasoline was still available for some vehicles, but now it has been phased out completely in the United States.

The lead ion, like most heavy metal ions, can permanently inactivate enzymes. Some consequences in children are lower birth weights, slower development of the neuromotor system, and diminished height. Symptoms of chronic lead poisoning appear when the blood carries only 25 μg/100 mL. They include headaches, loss of appetite, stomach aches, fatigue, and irritability, which in little children are not unusual complaints and normally do not arouse much alarm. In time, however, lead in circulation leads to incurable damage of the brain, liver, and kidneys.

The U.S. Centers for Disease Control estimates that 25% of all U.S. children have an excessive body burden of lead, defined as a level of over 25 μg/100 mL of a child's blood. Some specialists in heavy metal pollution at the U.S. Environmental Protection Agency, alarmed particularly by the effect of lead on nerve development, recommended that this level be reduced to 10 μg/100 mL, and in October of 1991 it was announced that the U.S. Centers for Disease Control would lower the allowed threshold. At a blood level of 10 μg of Pb/100 mL it is estimated that over 80% of all U.S. children and 77% of adults would carry an excessive burden.

Among little children growing up in tenements, the eating of flakes of lead-based paints was a bizarre cause of lead-based brain disease in the 1960s, until lead compounds were banned from interior paints.

One study on adult men found that blood pressure increases with increases in the levels of lead in the blood between 10 and 20 μg/100 mL.

Bismuth is one of only a few substances that expand slightly when they freeze. This property makes bismuth ideal for making accurate castings because it expands to fill all the details of the mold. The other principal use of bismuth is in making alloys with unusually low melting points. An example is Wood's metal, an alloy consisting of 50% bismuth, 25% lead, 12.5% tin, and 12.5% cadmium. This alloy has a melting point of only 70 °C; it melts when dipped into boiling water! Wood's metal is used to seal the heads of overhead sprinkler systems. A fire will trigger the system automatically by melting the alloy before the temperature has risen very high.

Sn [Kr] $4d^{10}5s^25p^2$
Sn^{2+} [Kr] $4d^{10}5s^2$
Sn^{4+} [Kr] $4d^{10}$

Oxidation States One of the major features of the chemistries of tin, lead, and bismuth is the occurrence of two oxidation states. Tin and lead have valence shell electron configurations corresponding to ns^2np^2 and form compounds in the +2 and +4 oxidation states. These correspond to the loss of first the pair of outer *p* electrons and then the pair of outer *s* electrons. Bismuth has the valence shell configuration $6s^26p^3$; its oxidation states are +3 and +5.

As with the elements in Group IIIA, the lower oxidation state becomes increasingly more stable relative to the higher one as we go down the group. Thus, for tin we find many compounds for both the +2 and +4 states, but the

+2 state is easily oxidized to the +4 state. On the other hand, lead in the +4 oxidation state is a powerful oxidizing agent, indicating the tendency of lead(IV) to be reduced to the more stable lead(II).

The relative stabilities of the oxidation states of tin and lead can also be seen in their reactions with acids. With a nonoxidizing acid, tin forms the +2 ion.

$$Sn(s) + 2HCl(aq) \longrightarrow SnCl_2(aq) + H_2(g)$$

But with nitric acid, an oxidizing acid, the product is tin(IV) oxide, SnO_2.

$$Sn(s) + 4HNO_3(aq) \longrightarrow SnO_2(s) + 4NO_2(g) + 2H_2O$$

Lead forms only the + 2 state, even when strong oxidizing acids attack it.

$$3Pb(s) + 8HNO_3(aq) \longrightarrow 3Pb(NO_3)_2(aq) + 2NO(g) + 4H_2O$$

The behavior of these two metals toward HNO_3 reflects the greater stability of lead(II) compared with tin(II).

Besides dissolving in acids, tin and lead also dissolve in base. The reactions are

SnO and PbO are amphoteric; they dissolve in acids (e.g., HNO_3) and bases (e.g., NaOH).

$$Sn(s) + 2OH^-(aq) + 2H_2O \longrightarrow \underset{\text{stannite ion}}{Sn(OH)_4^{2-}(aq)} + H_2(g)$$

$$Pb(s) + 2OH^-(aq) + 2H_2O \longrightarrow \underset{\text{plumbite ion}}{Pb(OH)_4^{2-}(aq)} + H_2(g)$$

The stannite ion is very easily oxidized and is therefore a strong reducing agent. The plumbite ion, on the other hand, is much more difficult to oxidize, which once more illustrates the relative stabilities of the oxidation states of these two metals.

For bismuth, the +3 state is much more stable than the +5 state. Oxidation using O_2 or Cl_2 gives only Bi_2O_3 and $BiCl_3$, for example. Oxidation of Bi_2O_3 to Bi_2O_5 is very difficult and requires a very strong oxidizing agent and special conditions. The compound $BiCl_5$ doesn't exist because bismuth(V) is such a strong oxidizing agent it would oxidize Cl^- to Cl_2.

Compounds of Tin, Lead, and Bismuth

Tin Tin forms halides in both its + 2 and + 4 oxidation states. Their general formulas are SnX_2 and SnX_4, where X stands for a halogen. One of the most important of these is tin(II) fluoride (stannous fluoride), SnF_2, an ingredient in some fluoride toothpastes. The tin(IV) halides are all covalent. An example is $SnCl_4$, a colorless liquid that freezes at − 33 °C and boils at 114 °C. The covalent nature of the Sn—Cl bonds in $SnCl_4$ can be explained in the same way as the covalent Be—Cl bonds in $BeCl_2$: The small size and high charge of a true Sn^{4+} ion would distort the electron cloud of the chloride ion so much that the electron density drawn between the two nuclei would constitute a covalent bond. Of course, we can also view the bonds in $SnCl_4$ as ordinary covalent bonds formed in the usual way.

The properties of $SnCl_4$ are those of a molecular substance, not an ionic one.

$$\cdot\dot{\underset{\cdot}{Sn}}\cdot + 4\cdot\ddot{\underset{\cdot\cdot}{Cl}}: \longrightarrow :\ddot{\underset{\cdot\cdot}{Cl}}-\underset{\underset{:\underset{\cdot\cdot}{Cl}:}{|}}{\overset{\overset{:\ddot{Cl}:}{|}}{Sn}}-\ddot{\underset{\cdot\cdot}{Cl}}:$$

In either case, the result is the same.

FIGURE 3.25

Red lead, Pb_3O_4.

The halides $PbCl_2$, $PbBr_2$, and PbI_2 are relatively insoluble in water and become less soluble going from the chloride to the iodide.

Pb_3O_4 can be formulated as $(PbO)_2 \cdot (PbO_2)$.

Today, TiO_2 has replaced white lead as a pigment in almost all paint formulations.

A drop of Na_2S solution placed on paint containing lead will immediately turn black. This is used as a test for lead-based paints in old buildings.

FIGURE 3.26

In an acidic solution, yellow $NaBiO_3$ oxidizes the nearly colorless Mn^{2+} ion to the violet-colored MnO_4^- ion.

The reaction of tin with oxygen gives tin(IV) oxide, SnO_2, rather than SnO. Tin(IV) oxide is used to give glass a transparent electrically conducting surface. Such conducting glass is used in electronic displays. Another oxygen-containing compound of tin is called bis(tributyltin) oxide. Its formula is $[(C_4H_9)_3Sn]_2O$. It is used in wood-treatment products that prevent rot. It has also been used in antifouling paints that are applied to boat hulls to prevent the growth of marine organisms such as barnacles. However, its toxicity to marine organisms has led to a ban on its use for this purpose.

Lead Lead forms halides in the +2 state, but in the +4 state only the chloride is relatively stable. Lead(IV) is such a powerful oxidizing agent that Br^- and I^- are oxidized by it, so $PbBr_4$ is very unstable and the existence of PbI_4 is doubtful. In fact, even $PbCl_4$ decomposes easily by an internal redox reaction.

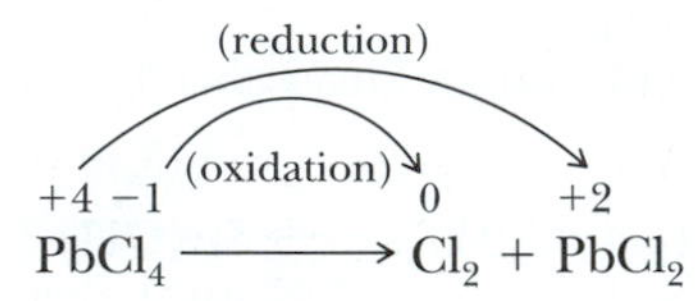

$$\overset{+4\ -1}{PbCl_4} \longrightarrow \overset{0}{Cl_2} + \overset{+2}{PbCl_2}$$

(reduction) (oxidation)

Among the more useful compounds of lead are the oxides. When heated in air, lead forms PbO—a yellow powder called *litharge* that is used in pottery glazes and in making fine lead crystal. If PbO is heated in air it can be oxidized further to give *red lead*—a mixed lead(II)–lead(IV) oxide with the formula Pb_3O_4 (Figure 3.25). This oxide is used in corrosion-inhibiting coatings applied to structural steel. Treatment of solutions of plumbite ion, $Pb(OH)_4^{2-}$, with hypochlorite ion produces yet another oxide, PbO_2. This compound is a strong oxidizing agent, and its most common use is as the cathode material in the lead storage battery.

For many years, a common pigment in paints was a white basic carbonate of lead, $Pb_3(OH)_2(CO_3)_2$, known generally as "white lead." Its use is now severely restricted because of the high toxicity of lead (as mentioned in Special Topic 3.2). As you probably know, children living in old buildings where there are remnants of former coats of lead-based paints often show signs of lead poisoning, presumably caused by eating chips of the lead-containing paint or by breathing dust contaminated by the lead-based paint. As a pigment, white lead has excellent covering power, but it is gradually darkened by exposure to air. This is caused by small amounts of hydrogen sulfide in the air formed by decomposition of organic matter. The H_2S reacts with the lead-based pigment forming black lead sulfide, PbS. Another lead-containing pigment used in artists' oil paints is lead chromate, $PbCrO_4$, which has a bright yellow color.

Bismuth Bismuth(III) fluoride is ionic, but its other halides, such as $BiCl_3$, are covalent. In water, $BiCl_3$ hydrolyzes (reacts with water) to give a precipitate having the formula BiOCl (named *bismuthyl chloride*).

$$BiCl_3 + H_2O \longrightarrow BiOCl(s) + 2H^+ + 2Cl^-$$

Other bismuth salts, such as its nitrate, $Bi(NO_3)_3$, and sulfate, $Bi_2(SO_4)_3$, also react with water to form the bismuthyl ion, BiO^+. In fact, a compound formed this way from the nitrate and having the approximate formula $BiO(NO_3)$ (the exact formula depends on how it is prepared) is called bismuth subnitrate and is used medicinally as an antacid in the treatment of gastric ulcers. Other bismuth compounds are also used in both the cosmetic and pharmaceutical industries and, combined, they account for approximately 30% of the bismuth produced each year.

When bismuth is heated in air, its oxide Bi_2O_3 is formed. As noted earlier, this can be further oxidized under extreme conditions to give bismuth in the +5 oxidation state. Compounds such as sodium bismuthate, $NaBiO_3$, which contain bismuth in the +5 state, are extremely powerful oxidizing agents. For example, the usual analytical test for manganese ion involves treating a solution made acidic with HNO_3 with $NaBiO_3$. Any Mn^{2+} in the solution is oxidized to permanganate ion, MnO_4^-, whose presence is easily detected because of its intense purple color (Figure 3.26). The reaction is

$$14H^+ + 5BiO_3^- + 2Mn^{2+} \longrightarrow 2MnO_4^- + 5Bi^{3+} + 7H_2O$$

3.6 TRANSITION METALS: GENERAL CHARACTERISTICS AND PERIODIC TRENDS

In the preceding sections we discussed the properties of the representative metals—those that belong to the A groups in the periodic table. Although some representative metals (beryllium, magnesium, aluminum, tin, and lead) have found practical uses, most of the metals we encounter day to day are found in the center of the periodic table.

All of the elements between Group IIA and Group IIIA of the periodic table are metals.[3] They are usually divided into two main categories. The **transition elements** or **transition metals** are the block of elements normally placed in the body of the table; they consist of the elements in the B groups plus Group VIII.[4] The **inner transition elements** are those in the two long rows that are usually found just below the main body of the table. The elements in the first of the long rows are called the **lanthanides** because they follow lanthanum ($Z = 57$). Those in the second long row are called the **actinides** because they follow actinium ($Z = 89$). The lanthanides and actinides are rare elements and their chemistries are not particularly important to us in our present studies, so we will have little more to say about them.

The lanthanides are also called *rare earth metals.*

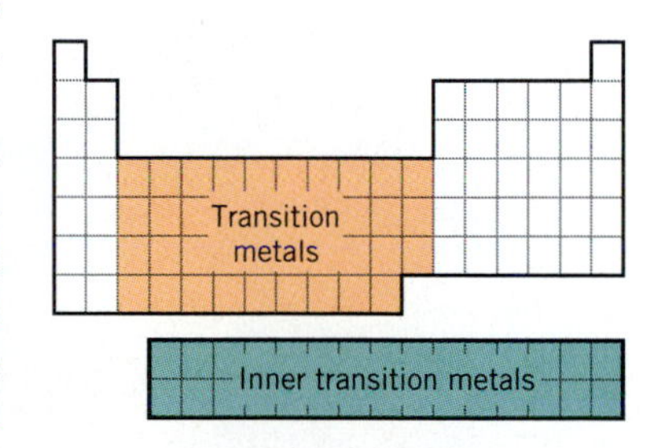

Like the other elements we have studied, the chemical and physical properties of the transition metals depend on their electron configurations. In Chapter 6 you saw that a *d* subshell is gradually filled as a period is crossed from left to right through a row of transition elements. Except for the metals in Group IIB, atoms of the transition elements all have partially filled *d* subshells. This is the main feature that distinguishes them from the representative elements.[5] In fact, the transition elements are often called the *d-block* elements.

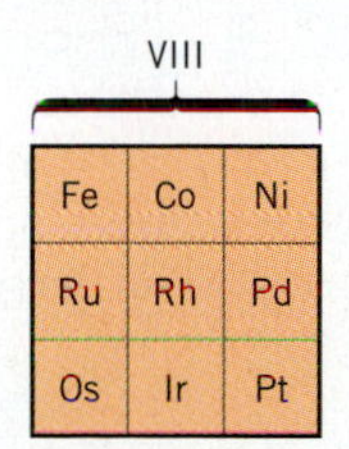

Characteristic Shared by Many of the Transition Elements

Many of the transition elements have properties in common. For example, they generally tend to be hard and have high melting points. This is particu-

[3] The B-group designations of the columns of transition elements originate from similarities with the A-group elements regarding the formulas of some compounds. For example, titanium in Group IVB forms TiO_2 just as carbon in Group IVA forms CO_2. However, even though many formulas are similar, the chemical and physical properties of the elements and their compounds are quite different.

[4] The elements collectively labeled Group VIII bear no similarities to the other elements in the periodic table in terms of the general formulas of their compounds. In addition, within Group VIII the horizontal similarities are greater than the vertical ones. The horizontal sets of elements are spoken of as *triads.* Iron, cobalt, and nickel constitute the *iron triad,* for example.

[5] The Group IIB elements have configurations corresponding to $(n - 1)d^{10}ns^2$ outside of a noble gas core, and some chemists prefer not to consider them true transition elements because their *d* subshells are complete.

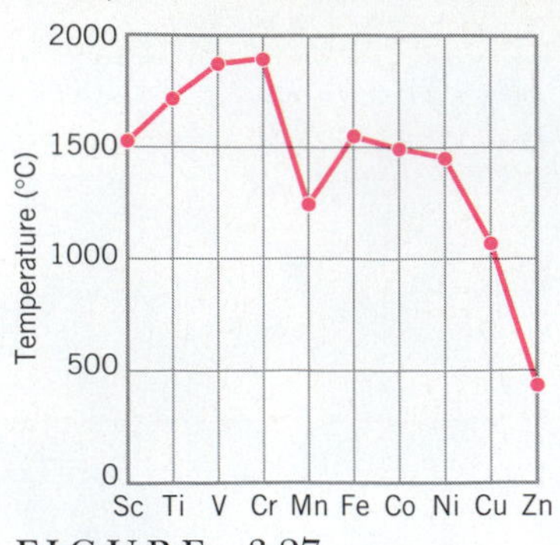

FIGURE 3.27

Melting points in °C of the period 4 transition elements.

larly true of the elements near the center of each row, as shown in Figure 3.27. The elements with the highest melting points also have the maximum number of unpaired *d* electrons, which suggests that the *d* electrons are probably involved to some extent in covalent bonding within the metallic lattice.

One of the most important characteristics of the transition metals is the occurrence of multiple oxidation states. The +2 state is common to many of these elements because many of their atoms have two electrons in their outermost *s* subshell. Examples are manganese, iron, and cobalt.

Mn	[Ar] $3d^5 4s^2$
Fe	[Ar] $3d^6 4s^2$
Co	[Ar] $3d^7 4s^2$

Each forms an ion by the loss of both 4*s* electrons (Mn^{2+}, Fe^{2+}, Co^{2+}). The underlying 3*d* subshell is fairly close in energy to the 4*s*, however, and not very much energy is needed to remove still another electron to give an ion with a 3+ charge.

Recall that complex ions are often simply called complexes.

Another property of the transition metals is the tendency of their ions to combine with neutral molecules or anions to form **complex ions** of the type discussed in Chapter 19. For example, you learned that the copper(II) ion combines with ammonia to form the deep blue complex $Cu(NH_3)_4{}^{2+}$.

Remember: The charge on a complex ion is the sum of the charges of the molecules and ions from which the complex is formed.

$$\underset{\text{pale blue}}{Cu^{2+}(aq)} + 4NH_3(aq) \longrightarrow \underset{\text{deep blue}}{Cu(NH_3)_4{}^{2+}(aq)}$$

The number of complexes formed by the transition elements is enormous, and their study is a major specialty of chemistry.

Many compounds and complexes of the transition metals have beautiful colors. For instance, all of the compounds of chromium are colored. In fact, chromium gets its name from the Greek word *chroma,* which means "color."

Variations in Atomic Size

In Chapter 6 you learned that among the *representative elements* the atomic radius increases from top to bottom within a group and decreases from left to right within a period. Among the *transition elements,* however, there are variations to these rules, which are understandable in terms of the locations of the elements within the periodic table.

As with the representative elements, there is a decrease in size among the transition elements as we move from left to right across a period. However, size changes more gradually for the transition elements and there is a size minimum near the center of each row, as shown in Figure 3.28.

The slowness of the size change is because as we go from one atom to the next the electrons that are being added enter a *d* subshell that is actually below the outer shell. Thus, for the transition elements in period 4, the outer electrons are in the 4*s* subshell, but it is the 3*d* subshell that is filled as we cross from left to right. These inner *d* electrons shield the outer electrons quite well from the increasing nuclear charge, so the *effective nuclear charge* felt by the outer electrons increases slowly, which leads to a slow decrease in size. The size minimum seems to occur because after the *d* subshell becomes more than half-filled, electron–electron repulsions force the orbitals in the *d* subshell to gradually expand in size, and this causes the sizes of the atoms to increase.

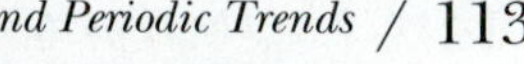

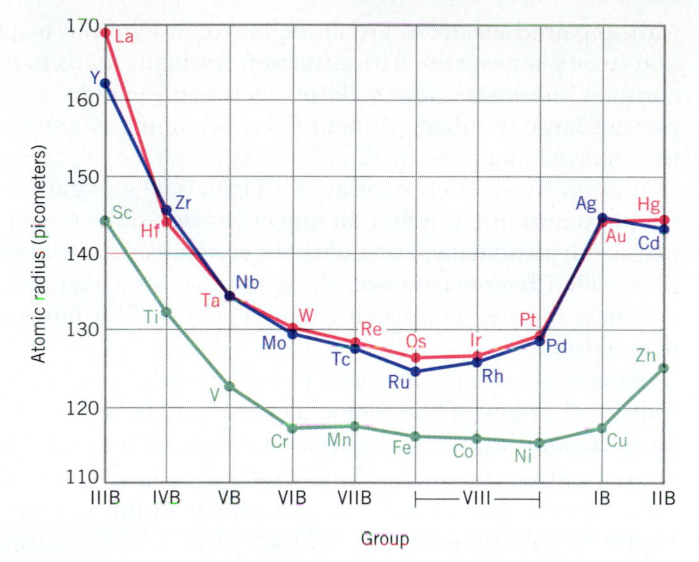

FIGURE 3.28

Size variations among the transition elements.

The Lanthanide Contraction When we turn our attention from periods to groups, we find the most unusual variations in atomic size. Normally we expect atomic size to increase going down a group, and we do find this for the elements in Group IIIB. However, for the other groups there is virtually no change in size going from period 5 to 6 as shown in Figure 3.28.

The reason for this phenomenon is the existence of the 14 elements of the lanthanide series that occur between lanthanum ($Z = 57$) and hafnium ($Z = 72$). As with any other row of elements, there is a gradual size decrease across the lanthanide series. This size decrease is sufficient to cause the elements that follow the lanthanides to be nearly the same size as the elements above them. In other words, for groups to the right of Group IIIB, the size *increase* we might expect going from periods 5 to 6 has been canceled by the size *decrease* across the lanthanide series. This phenomenon has been termed the **lanthanide contraction.**

Zirconium and hafnium occur together in nature because their ions have the same charge and the same size. These similarities make it difficult to separate them from each other.

Two major consequences of the lanthanide contraction are that the transition elements in period 6 are very dense and resistant to oxidation. They have unusually high densities because their atoms have nearly twice as much mass as do the atoms of the elements above them, but their sizes are virtually the same. Twice the mass occupying about the same volume leads to about twice the density.

The resistance of the period-6 transition metals to oxidation (loss of electrons) occurs because their outer electrons feel a very large effective nuclear charge. This means that the electrons are tightly held and are difficult to remove. The resistance of metals like platinum and gold to oxidation makes them commercially useful and adds to their overall value. Gold, for example, is used to plate electrical contacts in low-voltage circuits where even small amounts of corrosion would block the flow of current.

Magnetic Properties

Many of the atoms and ions of the transition elements contain unpaired electrons as a consequence of their partially filled *d* subshells. For instance, a manganese atom has five unpaired electrons, as does a Mn^{2+} ion. Substances with unpaired electrons are attracted to an external magnetic field and are said to be *paramagnetic*. The attraction normally tends to be weak because the constant movements and collisions between the individual atom-size magnets prevent large numbers of them from becoming aligned simultaneously with the external magnetic field.

A property we often associate with iron is its *strong* attraction to magnets. In fact, to determine whether an object is made of iron, you might test it with a magnet. Iron actually is one of three elements that exhibit this strong magnetism, called **ferromagnetism;** the other two are cobalt and nickel. Ferromagnetism is unusual because it is about one million times stronger than paramagnetism.

Ferromagnetism is also due to the presence of unpaired electrons in the individual atoms of the element. However, in iron, cobalt, and nickel, the paramagnetic atoms are spaced *just right* to interact with each other very strongly. Huge numbers of individual atomic magnets align and lock onto each other to give strong magnetic effects within regions called **domains.** As Figure 3.29 illustrates, in an ordinary piece of a ferromagnetic material these domains are randomly oriented, so the object does not appear to be a magnet. But in the magnetic field produced by another magnet, the domains shift and turn, and each time a domain turns a tremendous number of tiny magnets turn at once. Because so many ''atomic magnets'' line up with the field, a very strong attraction is felt. Furthermore, when the external magnetic field is removed, the domains don't immediately become random again. They stay aligned and the metal retains its magnetism; it has become a ''permanent magnet.'' It isn't really permanent, however. Heating it or pounding it with a hammer can provide the jostling needed to randomly orient the domains, and the ''permanent'' magnetism is lost.

Ferromagnetism is a property of the solid state. Melting the iron, cobalt, or nickel object also destroys the ferromagnetism because the domains of the solid state are lost when the metal is liquefied. The liquid is therefore just paramagnetic.

In some cases, it is possible to alter the spacings of the atoms in a paramagnetic solid to give it ferromagnetic properties. For example, pure manganese

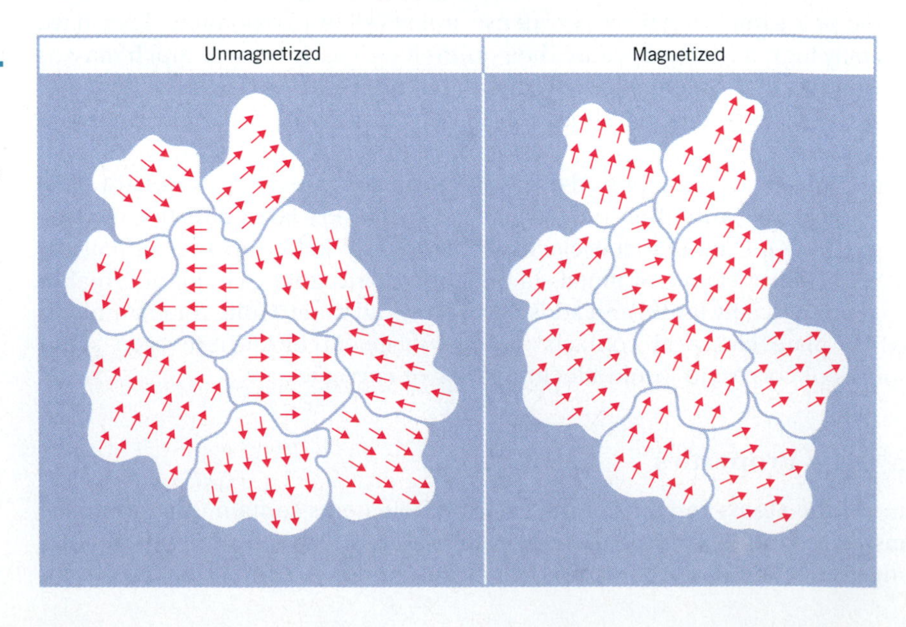

FIGURE 3.29

In a ferromagnetic solid, the individual paramagnetic atoms are aligned within domains. In the unmagnetized state, the domains are randomly oriented. When magnetized, the domains shift and become aligned. (From J. E. Brady and G. E. Humiston, *General Chemistry: Principles and Structure,* 4th ed., John Wiley and Sons, New York, 1986. Used by permission.)

is only paramagnetic, even though its atoms each contain five unpaired electrons. Adding copper to manganese changes the spacing between the manganese atoms, and if the proper amount of copper is added, a ferromagnetic alloy is formed.

Many ferromagnetic alloys have been made. Alnico magnets, for instance, are an alloy of iron, aluminum, nickel, copper, and cobalt.

3.7 PROPERTIES OF SOME IMPORTANT TRANSITION ELEMENTS

If you pause for a moment and look around, you will see a number of transition metals or their compounds. The paint on the wall probably contains titanium (as its oxide, TiO_2), the coins in your pocket or purse are copper or a copper–nickel alloy, and iron and steel alloys are surely nearby. Iron and many other transition metals are present in various quantities—some in only trace amounts—in your body. Thus, transition elements are clearly necessary for our comfort and even our existence. In this section we focus our attention on the specific properties of some of these important metals.

Titanium

Ti [Ar] $3d^2 4s^2$

Titanium is only about 60% as dense as iron, it is very strong, and it is resistant to corrosion. This combination of properties makes it a useful metal. It is used in place of steel and aluminum in aircraft because of its strength and low density, and it is used in jet engines because it doesn't lose its strength at high temperatures.

Titanium is found in several minerals, one of the most important of which is *rutile*, TiO_2. Separation of the metal from its ores is not easy, however, because titanium reacts directly with oxygen, nitrogen, and carbon at the temperatures that would be required for its reduction by ordinary metallurgical methods. Instead, the oxide is heated with carbon and chlorine, which converts it to the chloride, $TiCl_4$.

$$TiO_2 + C + 2Cl_2 \longrightarrow TiCl_4 + CO_2$$

Titanium tetrachloride is then reduced with magnesium under a blanket of the intert noble gas argon.

$$TiCl_4 + 2Mg \longrightarrow Ti + 2MgCl_2$$

Besides being involved in the production of titanium metal, $TiCl_4$ is one of the most important compounds of titanium because it is used to prepare most other compounds of this element. Titanium tetrachloride is a clear colorless liquid with a boiling point of only 136 °C and it is composed of covalently bonded molecules similar to $SnCl_4$. The U.S. Navy at one time used $TiCl_4$ to make smoke screens because it reacts almost instantly with moist air to form a dense fog of TiO_2 and HCl. (Figure 3.30.)

$$TiCl_4(g) + 2H_2O(g) \longrightarrow TiO_2(s) + 4HCl(g)$$

FIGURE 3.30
Titanium tetrachloride, $TiCl_4$, is a volatile liquid whose vapors react with moisture in the air to give a dense smoke of TiO_2. The reaction was once used by the U.S. Navy to make smoke screens during naval battles.

The high degree of covalence of $TiCl_4$ can be explained in the way we explained the covalence in $SnCl_4$: A small, highly charged Ti^{4+} ion placed next to a Cl^- ion would distort the anion's electron cloud so much that the electron density drawn between the two would constitute a covalent bond.

Although titanium occurs in several oxidation states, just like most of the other transition elements, by far the most stable is the +4 state. $TiCl_4$ is one example of a titanium(IV) compound. Another, and the most commercially important titanium compound, is titanium(IV) oxide, TiO_2. Usually it is called titanium dioxide, and is a brilliant white substance that is the most common white pigment in paint today. It is also used as a brightener in paper and as a sun screen.

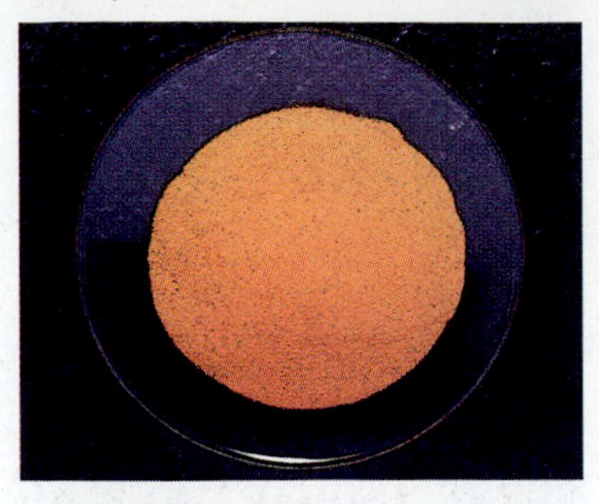

FIGURE 3.31

Vanadium(V) oxide, V_2O_5. The compound is also known as vanadium pentoxide.

Vanadium

V [Ar] $3d^34s^2$

Vanadium, like titanium, is difficult to extract from its compounds. Its chief use is in making special alloy steels and cast iron. Adding vanadium makes the alloys more ductile and shock resistant. Vanadium is also used in nuclear reactors because it is highly "transparent" to neutrons.

Vanadium also exhibits multiple oxidation states. Its highest is +5 and corresponds to the "loss" of its $4s$ and $3d$ electrons. The oxide V_2O_5, shown in Figure 3.31, is the catalyst in the contact process for the manufacture of sulfuric acid. V_2O_5 is a reasonably good oxidizing agent, and in Chapter 2 of this supplement we saw that it promotes the oxidation of SO_2 to SO_3 by O_2. Vanadium(V) oxide dissolves slightly in water to give acidic solutions, in acids to give the VO_2^+ ion, and in base to give the VO_4^{3-} ion.

Recall from Chapter 1 that the higher the oxidation state of the metal, the more acidic is its oxide.

The formation of oxocations such as VO_2^+ by metals in high oxidation states is not unusual. It can be viewed simply as a case of extreme hydrolysis of the cation. For the cation VO_2^+, we think of two water molecules that have been so extensively polarized by the positive charge of the V^{5+} cation that the hydrogens have come off as H^+ ions and become attached to water molecules to give H_3O^+ ions. (This explains the acidity of V_2O_5 solutions in water.) The reaction is diagrammed in Figure 3.32.

Vanadium also exists in compounds in the +4, +3, and +2 oxidation states, but they're of little practical importance.

Chromium

Cr [Ar] $3d^54s^1$

This metal is white, lustrous, hard, brittle, and very resistant to corrosion. These properties make it excellent as both a decorative and a protective coating over other metals such as brass, bronze, and steel. Almost everyone is familiar with the *chrome plate* that is deposited electrolytically on automobile or motorcycle parts. Large amounts of chromium are used to produce alloys. The best known is *stainless steel*—a type of steel alloy that is resistant to corrosion. A typical stainless steel contains about 18% chromium, 8% nickel, plus small amounts of manganese, carbon, phosphorus, sulfur, and silicon, all combined with iron. *Nichrome,* an alloy of chromium and nickel, is often used as the wire heating element in various heating devices such as toasters.

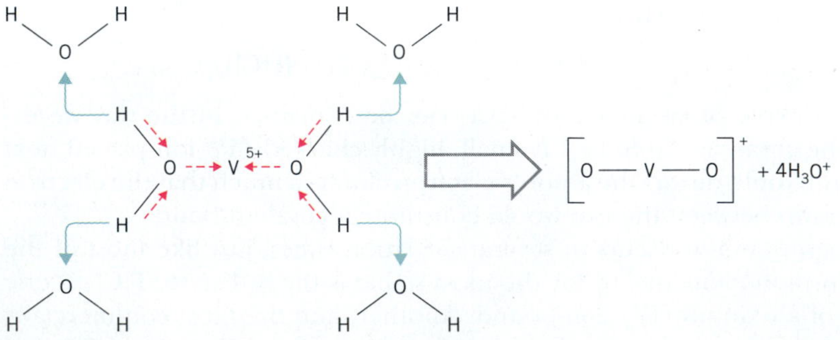

FIGURE 3.32

The formation of the VO_2^+ ion by extensive hydrolysis of the V^{5+} cation.

$$[O—V—O]^+ + 4H_3O^+$$

In compounds, chromium can exist in a number of different oxidation states; the most common are +2, +3, and +6. If elemental chromium is dissolved in dilute nonoxidizing acids that have been purged of oxygen, the pale blue chromium(II) ion or chromous ion, Cr^{2+}, is formed.

$$Cr(s) + 2H^+(aq) \longrightarrow Cr^{2+}(aq) + H_2(g)$$

This is the least stable oxidation state of chromium and it is very easily oxidized to the +3 state, even by molecular oxygen. (That's why the dilute acid must be oxygen free to give Cr^{2+}.)

Traces of O_2 can be removed from gases such as N_2 by bubbling them through a solution that contains Cr^{2+} ion.

The +3 state is the most stable for chromium. In water, Cr^{3+} actually exists as the blue-purple complex ion $Cr(H_2O)_6{}^{3+}$, and many chromium(III) salts owe their color to the presence of this ion in their crystals. An example is chrome alum, $KCr(SO_4)_2 \cdot 12H_2O$, shown in Figure 3.33. This double salt is formed when solutions that contain both K_2SO_4 and $Cr_2(SO_4)_3$ are gradually evaporated.

Cr^{3+} is known to form thousands of complexes.

FIGURE 3.33

Chrome alum has the formula $KCr(SO_4)_2 \cdot 12H_2O$. The solid and the solution contain the violet complex ion $Cr(H_2O)_6{}^{3+}$.

In some respects, chromium(III) and aluminum(III) are alike. Solutions that contain the $Cr(H_2O)_6{}^{3+}$ ion are slightly acidic, just as solutions that contain the $Al(H_2O)_6{}^{3+}$ ion.

$$Cr(H_2O)_6{}^{3+} + H_2O \rightleftharpoons Cr(H_2O)_5(OH)^{2+} + H_3O^+$$

In the +3 state, chromium is also amphoteric. Adding base to a solution containing $Cr(H_2O)_6{}^{3+}$ precipitates the pale blue-violet gelatinous hydroxide (Figure 3.34). Making the solution more basic causes the precipitate to dissolve, yielding a deep green solution.

$$\underset{\text{blue-purple}}{Cr(H_2O)_6{}^{3+}(aq)} + 3OH^-(aq) \longrightarrow \underset{\text{blue-violet}}{Cr(H_2O)_3(OH)_3(s)} + 3H_2O$$

$$\underset{\text{blue-violet}}{Cr(H_2O)_3(OH)_3(s)} + OH^-(aq) \longrightarrow \underset{\text{green}}{Cr(H_2O)_2(OH)_4{}^-(aq)} + H_2O$$

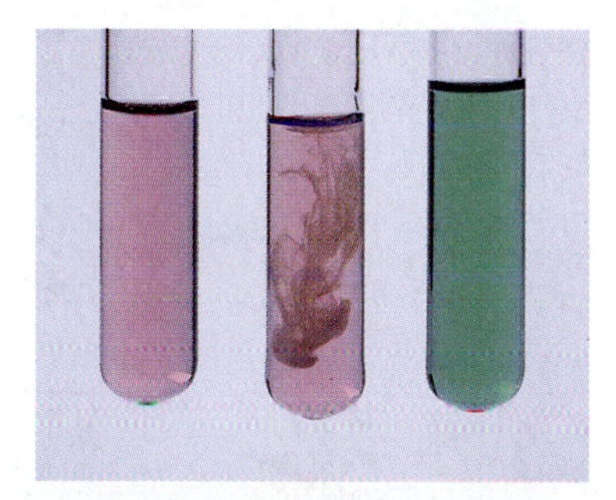

FIGURE 3.34

On the left, a solution containing $Cr(H_2O)_6{}^{3+}$. In the center, the precipitate of $Cr(H_2O)_3(OH)_3$ that is formed when the first solution is made slightly basic. When more base is added (right), the precipitate dissolves and yields a green solution of $Cr(H_2O)_2(OH)_4{}^-$.

The third important oxidation state of chromium is +6. The highly poisonous red-orange oxide CrO_3 (Figure 3.35) is a powerful oxidizing agent and is the acid anhydride of chromic acid, H_2CrO_4. Chromic acid is the primary species in aqueous solutions at very low pH (i.e., in very acidic solutions). At higher pH, two other species predominate, the yellow *chromate ion,* $CrO_4{}^{2-}$, and the red-orange *dichromate ion,* $Cr_2O_7{}^{2-}$ (Figure 3.36). There is an equilibrium between $CrO_4{}^{2-}$ and $Cr_2O_7{}^{2-}$ that can be written

$$2CrO_4{}^{2-}(aq) + 2H^+(aq) \rightleftharpoons Cr_2O_7{}^{2-}(aq) + H_2O$$

In solutions that are acidic, this equilibrium is shifted to the right and dichromate predominates. The principal ion present in basic solutions is $CrO_4{}^{2-}$ because the reaction is shifted to the left as the H^+ concentration decreases.

The formation of $Cr_2O_7{}^{2-}$ from $CrO_4{}^{2-}$ is easier to understand if the equilibrium is written

$$2HCrO_4{}^-(aq) \rightleftharpoons H_2O + Cr_2O_7{}^{2-}(aq)$$

As the pH is lowered and the hydrogen ion concentration increases, more and more $HCrO_4{}^-$ is formed in the solution by the addition of protons to $CrO_4{}^{2-}$ ions. Using Lewis structures, we see that removal of the components of water from a pair of $HCrO_4{}^-$ ions joins the chromium atoms with an oxygen bridge.

FIGURE 3.35
Chromium(VI) oxide, CrO_3.

$$\left(\begin{array}{c} :\ddot{O}: \\ | \\ :\ddot{O}-Cr-\ddot{O}: \\ | \\ :\ddot{O}: \\ | \\ H \end{array}\right)^{-} + \left(\begin{array}{c} H \\ | \\ :\ddot{O}: \\ | \\ :\ddot{O}-Cr-\ddot{O}: \\ | \\ :\ddot{O}: \end{array}\right)^{-} \rightleftharpoons \left(\begin{array}{c} :\ddot{O}: \\ | \\ :\ddot{O}-Cr-\ddot{O}: \\ | \\ :O: \\ | \\ :\ddot{O}-Cr-\ddot{O}: \\ | \\ :\ddot{O}: \end{array}\right)^{2-} + H_2O$$

All of the chromium(VI) species are good oxidizing agents, although their strongest oxidizing abilities occur in acidic solutions.

Compounds of chromium find many practical applications. One of the principal uses is in pigments. For example, the oxide Cr_2O_3 shown in Figure 3.37 is the most stable green pigment known; it is used for coloring paints, roofing granules, cements, and plaster. Another pigment called *zinc yellow* (actually zinc chromate, $ZnCrO_4$) is used as a corrosion-inhibiting primer on aluminum and magnesium aircraft parts. As mentioned earlier, lead chromate is a yellow pigment used in artists' oil colors. Barium chromate, also insoluble in water, is another yellow pigment.

Pigments account for about 35% of the chromium chemicals that are produced annually. Another 25%, principally in the form of $Cr_2(SO_4)_3$, is used in tanning leather. Much of the rest is used in treating metal surfaces for corrosion control.

FIGURE 3.36
The solution on the left contains the red-orange dichromate ion, $Cr_2O_7^{2-}$; the one on the right contains the yellow chromate ion, CrO_4^{2-}.

Manganese

Mn [Ar] $3d^54s^2$

Manganese has many properties that are similar to those of iron. It corrodes in moist air, for example, and it dissolves in dilute acids with the evolution of hydrogen.

$$Mn(s) + 2H^+(aq) \longrightarrow Mn^{2+}(aq) + H_2(g)$$

The metal's chief uses are as an additive to steel and in the preparation of other alloys such as *manganese bronze* (a copper–manganese alloy) and *manganin* (an alloy of copper, manganese, and nickel whose electrical resistance changes only slightly with temperature).

The most important oxidation states of manganese are +2, +3, +4, +6, and +7. The most stable is the +2 state, which is formed by the removal of the outer 4*s* electrons from the manganese atom.

$$Mn\ [Ar]\ 3d^54s^2 \longrightarrow Mn^{2+}\ [Ar]\ 3d^5 + 2e^-$$

The Mn^{2+} ion thus has a half-filled *d* subshell—a configuration you may recall that is particularly stable. In water, the ion exists as the pale pink complex $Mn(H_2O)_6^{2+}$, and this is the color of most manganese(II) salts, because they usually crystallize from solutions as hydrates that contain this ion. An example is manganese chloride, shown in Figure 3.38.

FIGURE 3.37
Chromium(III) oxide, Cr_2O_3, is a stable green pigment that has many uses.

When made basic, Mn^{2+} precipitates as pale pink $Mn(OH)_2$. This hydroxide is easily oxidized in air to the compound $MnO(OH)$, which contains manganese in the + 3 state. Here again we have a compound in which H^+ has been lost from H_2O or OH^- to give an oxygen firmly attached to a metal ion, a phenomenon that we noted for vanadium(V) in water, where it exists as VO_2^+.

The least stable oxidation state of manganese is +7, which has a strong tendency to be reduced and is therefore a powerful oxidizing agent. The most common compound of manganese(VII) is *potassium permanganate,* $KMnO_4$, which dissolves in water to give deep purple solutions that contain the MnO_4^- ion. In an acidic solution, MnO_4^- is usually reduced to Mn^{2+}, which has a very pale pink color that is practically invisible if the solution is dilute.

$$\underset{\text{deep purple}}{MnO_4^-(aq)} + 8H^+(aq) + 5e^- \longrightarrow \underset{\text{almost colorless}}{Mn^{2+}(aq)} + 4H_2O$$

FIGURE 3.38

Crystals and solutions of $MnCl_2 \cdot 6H_2O$ are pink because they contain the complex ion $Mn(H_2O)_6^{2+}$.

MnO_2 is the cathode reactant in dry cells.

As we've noted earlier, many analytical procedures use $KMnO_4$ as an oxidizing agent in titrations where it serves as its own indicator.

When permanganate ion is reduced in neutral or basic solutions, reduction ceases at the + 4 state with the formation of manganese(IV) oxide, MnO_2 (better known as *manganese dioxide*).

$$MnO_4^-(aq) + 2H_2O + 3e^- \longrightarrow MnO_2(s) + 4OH^-(aq)$$

Manganese dioxide is a common compound from which many other manganese compounds are made. It tends to be **nonstoichiometric,** which means that the ratio of oxygen to manganese is somewhat variable and is not exactly 2 to 1 as indicated by its formula. Manganese dioxide is found in the manganese nodules scattered on the ocean floor, which are described above in Section 3.2. According to one theory, microorganisms may have extracted manganese from the seawater and deposited it as MnO_2 in the nodules.

When MnO_2 is added to molten potassium hydroxide and oxidized with air or potassium nitrate, the green *manganate ion,* MnO_4^{2-}, is formed. This ion is stable only in very basic solutions. When acidified, a portion of the MnO_4^{2-} is oxidized while the rest is reduced. The products are MnO_4^- and MnO_2.

$$3MnO_4^{2-}(aq) + 4H^+(aq) \longrightarrow 2MnO_4^-(aq) + MnO_2(s) + 2H_2O$$

A solution that contains the violet MnO_4^- ion.

Oxidation of water produces O_2:

$2H_2O \longrightarrow O_2 + 4H^+ + 4e^-$

Recall that when one portion of a chemical is oxidized by the rest, the reaction is called *disproportionation.* We say the MnO_4^{2-} disproportionates.

When manganese metal is oxidized in air, the final product is the oxide Mn_2O_3, which contains manganese in the +3 oxidation state. The Mn^{3+} ion is a fairly strong oxidizing agent and slowly oxidizes H_2O to O_2. It can be stabilized, however, by forming complex ions, and many complexes of manganese(III) are known.

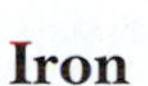

Iron

Fe [Ar] $3d^64s^2$

You are probably more familiar with this metal than with any other. It is relatively inexpensive, and iron and its alloys have such useful physical properties that they have been put to more uses than any other metal.

Iron is the second most abundant metal, next to aluminum, and it is the fourth most abundant element in the Earth's crust. The molten core of the earth is thought to be composed mostly of iron and nickel. In the pure state, iron is white and lustrous, but it is not especially hard. It is also quite reactive, being attacked by nonoxidizing acids such as HCl or H_2SO_4 to generate hydrogen.

The name *hematite* suggests the almost blood-red color of some of these iron ores.

$$Fe(s) + 2H^+(aq) \longrightarrow Fe^{2+}(aq) + H_2(g)$$

Oddly enough, iron doesn't react with concentrated nitric acid. Instead, its surface is made quite unreactive; the iron is said to have been made *passive.* Presumably, this is caused by the formation of an adhering oxide coating produced by the strongly oxidizing HNO_3.

The production and metallurgy of iron were discussed in Section 3.2. Its chief ores have traditionally been the red-orange oxide *hematite,* Fe_2O_3, and black *magnetite,* Fe_3O_4. In the United States, rich sources of these ores have been largely depleted through mining operations like that shown in Figure 3.39. A rock called *taconite,* consisting of fine crystals of magnetite bound together by silicate minerals, is now the focus of attention.

FIGURE 3.39

Open pit iron mining like that shown here has largely depleted deposits of hematite (Fe_2O_3) ore in the United States. Mining operations today focus on processing a rock-hard ore called taconite, which contains magnetic Fe_3O_4.

A problem that has plagued civilization ever since iron was discovered is its reaction with moisture and air. The corrosion product is rust, a hydrated oxide whose formula is usually given as $Fe_2O_3 \cdot xH_2O$. It doesn't adhere well to the metal, but instead falls away, exposing fresh iron to attack. The way that the corrosion of iron occurs and how it can be prevented are discussed in Special Topic 18.1 (page 782).

Iron forms compounds in two principal oxidation states, +2 and +3. Both states are relatively stable, both in the solid state in compounds and in aqueous solutions. Unlike the transition metals that appear earlier in period 4, iron forms no compounds in an oxidation state that would, in principle, involve all eight of its $3d$ and $4s$ electrons. The highest oxidation state observed for iron is +6, and that state is rare (and, therefore, unimportant to us here). This behavior of iron illustrates a general trend:

> The lower oxidation states of the transition metals become progressively more stable relative to the higher ones as we move from left to right across a period.

Iron forms compounds with most of the nonmetals. For example, it reacts with oxygen, chlorine, sulfur, phosphorus, and carbon when heated. The oxides of iron are FeO, Fe_2O_3, and Fe_3O_4. Crystalline FeO is difficult to prepare. When heated, it undergoes disproportionation to give both Fe and Fe_2O_3. The most common oxide is Fe_2O_3, which is the one found in rust and the ore hematite. As mentioned above, Fe_3O_4 occurs as the mineral magnetite, and as its name suggests, it is magnetic. It contains iron in both the +2 and +3 oxidation states.

FIGURE 3.40

Solutions that contain iron(III), like that on the left, have a yellow color from the hydrolysis of the $Fe(H_2O)_6^{3+}$ ion. Solutions of iron(II) salts (right) have a pale greenish-blue color caused by the ion $Fe(H_2O)_6^{2+}$.

In water, the chemistry of the Fe^{2+} and Fe^{3+} ions is largely the chemistry of their complex ions (Figure 3.40). For example, in water the iron(II) ion has a pale bluish-green color, which is the color of the $Fe(H_2O)_6^{2+}$ complex ion. Solid salts that contain Fe^{2+} are often pale green because this complex persists in the crystals. In water, iron(III) salts tend to be somewhat acidic and often appear slightly yellow because of hydrolysis of the $Fe(H_2O)_6^{3+}$ ion. The equilibria here are similar to those we described earlier in this chapter to explain the acidity of solutions of Cr^{3+} and Al^{3+}.

$$Fe(H_2O)_6^{3+}(aq) + H_2O \rightleftharpoons Fe(H_2O)_5(OH)^{2+}(aq) + H_3O^+(aq)$$

The reaction has an equilibrium constant, $K_a = 9 \times 10^{-4}$, which makes the $Fe(H_2O)_6^{3+}$ ion a stronger acid than acetic acid!

Addition of base to a solution that contains iron(II) ion precipitates a nearly colorless $Fe(OH)_2$, but if exposed to air the moist precipitate rapidly turns brown as it is oxidized by oxygen to give brown $Fe_2O_3 \cdot xH_2O$. Addition of base to a solution that contains iron(III) ion produces a reddish-brown gelatinous hydroxide, but unlike those of aluminum and chromium, it is not amphoteric; it does not dissolve in excess base (Figure 3.41).

$$Fe(H_2O)_6^{3+}(aq) + 3OH^-(aq) \longrightarrow Fe(H_2O)_3(OH)_3(s) + 3H_2O$$

In writing their formulas, gelatinous metal hydroxides such as this are often given in a simplified form: $Fe(OH)_3$.

Actually, the composition of the precipitate isn't as clearly defined as the equation above might suggest, and its formula is usually just given as the hydrated oxide we have seen earlier, $Fe_2O_3 \cdot xH_2O$.

Iron ions form many complex ions. Addition of thiocyanate ion, SCN^-, to a solution containing Fe^{3+} produces the blood-red complex $Fe(SCN)_6^{3-}$. The color of this complex is so intense that its formation serves as a qualitative test for the presence of Fe^{3+} in a solution.

Among the more interesting compounds of iron are those involving its cyanide complexes, whose common names and formulas are *ferrocyanide ion,* $Fe(CN)_6^{4-}$, and *ferricyanide ion,* $Fe(CN)_6^{3-}$. Taking the charge of the cyanide ion to be 1−, you can see that the ferrocyanide ion contains iron(II) (ferrous) and the ferricyanide ion contains iron(III) (ferric).

$Fe(CN)_6^{4-}$ is the hexacyanoferrate(II) ion.
$Fe(CN)_6^{3-}$ is the hexacyanoferrate(III) ion.

If a solution that contains Fe^{3+} is added to a solution containing $Fe(CN)_6^{4-}$, a blue precipitate known as *Prussian blue* is formed, as shown in Figure 3.42. Evidence suggests that its formula is $Fe_4[Fe(CN)_6]_3 \cdot 16H_2O$. Interestingly, the exact same compound is formed if Fe^{2+} ion is added to a solution of $Fe(CN)_6^{3-}$. The intense blue color of this precipitate is the basis of the blueprint process, once used by engineers and architects to copy their plans and drawings. Blueprint paper is made by treating paper with solutions of ammonium ferricyanide, $(NH_4)_3Fe(CN)_6$, and iron(III) citrate. When the paper is exposed to light, the Fe^{3+} ion of the citrate salt is reduced by the citrate ion to Fe^{2+}, and then the Fe^{2+} reacts with the $Fe(CN)_6^{3-}$ to give the deep blue precipitate. Finally, the unreacted ammonium ferricyanide and iron(III) citrate are washed from the paper so no further reaction can occur when the blueprints are used out in the field.

$Fe_4[Fe(CN)_6]_3 \cdot 16H_2O$ contains both iron(II) and iron(III).

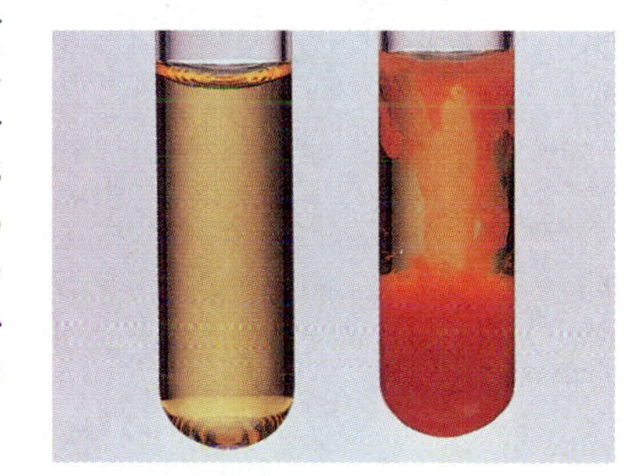

FIGURE 3.41

On the left, a solution that contains $Fe(H_2O)_6^{3+}$. When base is added, a reddish-brown gelatinous precipitate of $Fe(H_2O)_3(OH)_3$ is formed.

Cobalt

Co [Ar] $3d^74s^2$

Cobalt is a hard, bluish-white metal that is used mostly in catalysts and alloys. For example, it is combined with chromium and tungsten in an alloy called *stellite,* which retains its hardness even when hot. This property makes stellite useful for high-speed cutting tools (e.g., drill bits) used to machine steel. As noted earlier, cobalt is also used to prepare an alloy called *alnico,* which forms powerful magnets.

Chemically, cobalt is somewhat less active than iron, although it dissolves slowly in nonoxidizing acids such as HCl to liberate hydrogen.

$$Co(s) + 2H^+(aq) \longrightarrow Co^{2+}(aq) + H_2(g)$$

There are two important oxidation states of cobalt, +2 and +3. In water, the most stable is Co^{2+}, which exists as the deep-pink $Co(H_2O)_6^{2+}$ complex. This ion persists in hydrated cobalt(II) salts such as red crystals of cobalt chloride hexahydrate. On the labels of bottles of this chemical purchased from chemical supply firms, the formula is written $CoCl_2 \cdot 6H_2O$, but the chemical is really $[Co(H_2O)_6]Cl_2$. This particular salt is interesting because if it is heated, it loses water and turns blue, as shown in Figure 3.43. It is believed that the nature of the complex changes, as illustrated by the equation

A photograph showing the color of the $Co(H_2O)_6^{2+}$ ion and a solid containing this complex is on page 815 of the text.

$$\underset{\text{pink}}{[Co(H_2O)_6]Cl_2(s)} \xrightleftharpoons{\text{heat}} \underset{\text{blue}}{[Co(H_2O)_4]Cl_2(s)} + 2H_2O(g)$$

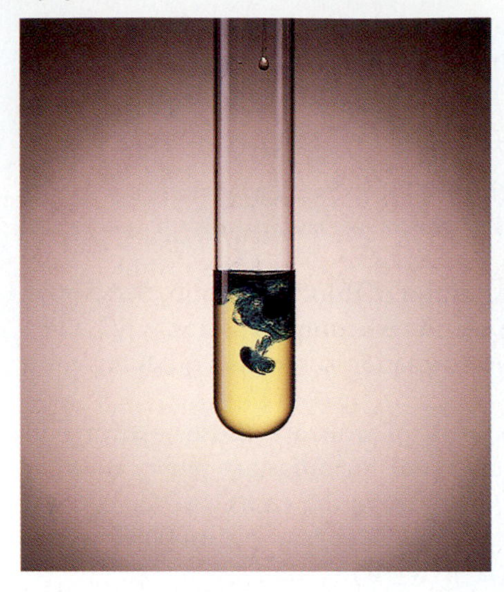

FIGURE 3.42

At the left we see the deep blue color of *Prussian blue,* $Fe_4[Fe(CN)_6]_3 \cdot 16H_2O$, formed when a drop of a dilute solution containing Fe^{3+} is added to a dilute solution of ferrocyanide ion, $Fe(CN)_6^{4-}$. On the right an architect studies blueprints of plans for the alteration of a building. The blue dye in the paper is $Fe_4[Fe(CN)_6]_3 \cdot 16H_2O$.

The change is reversible; when the blue crystals are exposed to moisture, water is recaptured and the solid becomes pink again. (This color change can be used to detect moisture.)

Cobalt(III) ion is a strong enough oxidizing agent to oxidize water unless the ion is bound in a complex. Many Co^{3+} complexes have been studied, and some of them are examined in Chapter 19.

Nickel

Ni [Ar] $3d^84s^2$

Nickel is one of our most useful metals. In the pure state, it resists corrosion, and metals such as iron and steel are frequently given thin protective coatings

(*a*)

(*b*)

FIGURE 3.43

(*a*) Pink hydrated $CoCl_2 \cdot 6H_2O$, which actually contains $[Co(H_2O)_6]Cl_2$. (*b*) After heating and driving off some of the water, blue $[Co(H_2O)_4]Cl_2$ remains.

of nickel by electrolysis. When alloyed with iron, nickel makes the metal more ductile and resistant to corrosion, and we mentioned earlier that nickel and chromium are the chief additives to iron in making stainless steel. Nickel also makes steel resistant to impact, a particularly desirable property in armor plate. Combined with copper, nickel produces a hard and strong alloy called *monel* that resists corrosion. Propeller shafts made of monel are valued in boats that operate in the corrosive environment of seawater. Nickel is also used as a catalyst for hydrogenation of organic compounds that contain double bonds.

The "nickel" coin is a copper–nickel alloy.

When recovered from its ores, nickel is only about 96% pure and must be refined. This can be done electrolytically as in the refining of copper described in Chapter 18. Another method for purifying nickel is called the *Mond process.* The impure nickel is warmed in a stream of carbon monoxide gas, where it forms a compound called *nickel carbonyl,* $Ni(CO)_4$. This compound is interesting because it contains nickel in the zero oxidation state. Nickel carbonyl is a liquid with a high vapor pressure, so as it is formed its vapor is carried away by the stream of carbon monoxide, leaving impurities behind. The $Ni(CO)_4$ vapor is then passed over a very hot surface on which it decomposes into pure metallic nickel and carbon monoxide. The CO is then recycled. The Mond process produces nickel that's about 99.95% pure.

$Ni(CO)_4$ is very toxic.

Nickel is a moderately active metal, and it dissolves in nonoxidizing acids such as HCl to give Ni^{2+} and H_2.

Nickel compounds are added to glass to give it a green color.

$$Ni(s) + 2H^+(aq) \longrightarrow Ni^{2+}(aq) + H_2(g)$$

As with many of the other transition metals we've studied, the nickel(II) ion exists in water as a complex ion; its formula is $Ni(H_2O)_6{}^{2+}$ and it has an emerald green color (Figure 3.44). Many crystalline nickel salts are "hydrates" and are green because the crystals contain this complex. Nickel forms many complexes in its +2 oxidation state. Another is the blue $Ni(NH_3)_6{}^{2+}$ ion that is formed when ammonia is added to a solution that contains Ni^{2+}.

The +2 oxidation state is the most stable one for nickel, and this is how we find it in nearly all its compounds. Higher oxidation states are powerful oxidizing agents because they have a strong tendency to be reduced to the +2 state. You may recall that the nickel–cadmium cell described in Chapter 18 uses nickel(IV) oxide (NiO_2) as the cathode. The cell reaction during discharge is

$$Cd(s) + NiO_2(s) + 2H_2O \longrightarrow Cd(OH)_2(s) + Ni(OH)_2(s)$$

FIGURE 3.44

The complex ion $Ni(H_2O)_6{}^{2+}$ gives its green color to both crystals and solutions of $NiCl_2 \cdot 6H_2O$.

Copper, Silver, and Gold—The Coinage Metals

Copper, silver, and gold are often called the **coinage metals** because they have been used for that purpose since ancient times. Each has an outer electron configuration of $(n-1)d^{10}ns^1$. The loss of the single *s* electron gives the +1 oxidation state, which is the reason for the IB designation for the group. In the case of copper, a second electron can be lost easily, and most copper compounds contain Cu^{2+}. Gold loses another two electrons rather easily, so it tends to form Au^{3+} rather than Au^+.

Cu $3d^{10}4s^1$
Ag $4d^{10}5s^1$
Au $5d^{10}6s^1$

All the Group IB metals are commercially valualbe. Copper is a metal that every U.S. citizen would recognize because it has been used for years to make the penny (although, since 1981, new pennies have been made of zinc with just a thin copper coating). Copper has a very high electrical and thermal conductivity, and for this reason it is used in electrical wiring. In fact, the largest concentration of copper in the world is said to be the copper wire that

Free elemental copper, called *native copper,* is found in the upper part of Michigan.

In recent years, pennies have been made of zinc, with only a thin coating of copper. Here we see a new penny that has been dipped in nitric acid, which has dissolved this copper coating to expose the gray zinc beneath.

lies beneath the streets of New York City. Copper is also fairly resistant to corrosion and is widely used as pipe to carry hot and cold water in homes and office buildings.

Silver has the highest thermal and electrical conductivity of any metal. Its value as a coinage metal, however, makes it too expensive to be used often as an electrical conductor. Silver has a high luster and, when polished, reflects light very well. This makes it valuable for jewelry and for the reflecting coating on mirrors. Silver is a soft metal and is usually alloyed with copper. *Sterling silver,* for example, contains 7.5% copper, and silver used in jewelry often contains up to 20% copper. One of silver's most important applications is in photography. Silver compounds tend to be unstable and sensitive to light (see Special Topic 3.3).

Everyone knows of gold's value as bullion and as a decorative metal for jewelry. As mentioned earlier, gold is also used occasionally to plate electrical contacts because of its low chemical reactivity. Pure gold is very soft and is particularly ductile and malleable. Gold leaf, used for decorative lettering in signs, is made by pounding gold into very thin sheets. It is so thin that some light passes through it. A stack of 11,000 gold leaflets is only 1 mm thick.

Copper is often found as Cu_2S in ores. Their treatment is discussed in Section 3.2 of this supplement.

Chemical Properties of Copper The elements of Group IB are not as reactive as other metals that we've discussed so far. Copper, for example, isn't attacked by nonoxidizing acids such as HCl. As we've noted on several occasions, however, it does dissolve in both dilute and concentrated nitric acid. (See page 380.) Copper also is attacked by hot concentrated sulfuric acid, which is a fairly potent oxidizing agent.

$$Cu(s) + 2H_2SO_4(\text{concd}) \xrightarrow{\text{heat}} CuSO_4(s) + SO_2(g) + 2H_2O$$

Copper sulfate, the product of this reaction, is one of the more important compounds of copper. We usually encounter it as the blue hydrate $CuSO_4 \cdot 5H_2O$. The dehydration of this compound, you may recall, was discussed in Chapter 2 of the text (page 41).

When heated strongly in air, copper acquires a coating of its black oxide, CuO. Although fairly resistant to corrosion, copper does oxidize slowly in air, and when CO_2 is also present its surface becomes coated with a green film of $Cu_2(OH)_2CO_3$. The outer surface of the Statue of Liberty is made of copper, and this compound is what gives it its green color (Figure 3.45).

FIGURE 3.45
The green color of the Statue of Liberty is caused by a surface coating of $Cu_2(OH)_2CO_3$ that covers the copper skin of the statue.

As mentioned above, copper forms compounds in two oxidation states, +1 and +2. The Cu^+ ion is unstable in water and disproportionates.

$$2Cu^+(aq) \longrightarrow Cu(s) + Cu^{2+}(aq)$$

However, it can be isolated in compounds such as CuCl, which is insoluble. For example, heating a solution that contains $CuCl_2$ and metallic copper acidified by HCl produces CuCl.

$$Cu(s) + Cu^{2+}(aq) + 2Cl^-(aq) \longrightarrow 2CuCl(s)$$

One of the water molecules in $CuSO_4 \cdot 5H_2O$ is hydrogen bonded to the sulfate ion.

The most stable oxidation state of copper is +2, and most copper compounds contain the Cu^{2+} ion, often in the form of a complex ion such as the blue $Cu(H_2O)_4{}^{2+}$ complex. This is what gives the color to crystals of $CuSO_4 \cdot 5H_2O$, and it is the ion present in dilute solutions of most copper salts. Copper(II) ion forms many complexes, such as the deep-blue $Cu(NH_3)_4{}^{2+}$

ion formed when aqueous ammonia is added to a solution that contains the $Cu(H_2O)_4^{2+}$. Concentrated solutions of the salt $CuCl_2$ are green because they contain a mixture of complexes such as yellow $CuCl_4^{2-}$ and blu$Cu(H_2O)_4^{2+}$. And adding base to a solution of a soluble copper(II) salt first precipitates $Cu(OH)_2$ and then redissolves the precipitate to give the complex $Cu(OH)_4^{2-}$ (Figure 3.46).

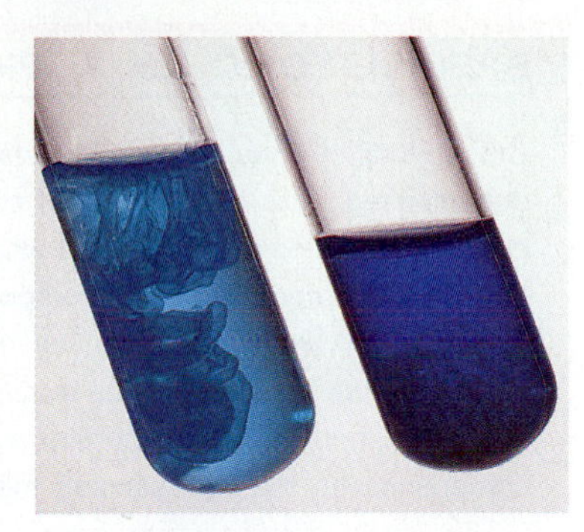

FIGURE 3.46

On the left, NaOH precipitates blue $Cu(OH)_2$. Further addition of NaOH (right) has caused $Cu(OH)_2$ to dissolve, forming the complex ion $Cu(OH)_4^{2-}$.

Chemical Properties of Silver Silver is even more difficult to oxidize than copper, so it isn't attacked by nonoxidizing acids either. But, like copper, silver does dissolve in concentrated and dilute nitric acid, yielding NO_2 and NO as reduction products, respectively, along with one of silver's most important water-soluble compounds, silver nitrate ($AgNO_3$). Metallic silver isn't attacked by oxygen of the air, but it does tarnish in air by reacting with oxygen and traces of H_2S (formed in nature by decomposing vegetation). The black tarnish deposit is silver sulfide.

$$4Ag(s) + 2H_2S(g) + O_2(g) \longrightarrow 2Ag_2S(s) + 2H_2O$$

Similar reactions occur if silver utensils are left in contact with sulfur-containing foods such as eggs or mustard.

The most important oxidation state of silver is +1. Many compounds of Ag^+ are insoluble in water. For example, adding a base to a solution of Ag^+ yields a brown precipitate of silver oxide.

Silver is often found free in nature. It also occurs as AgCl and Ag_2S.

$$2Ag^+(aq) + 2OH^-(aq) \longrightarrow Ag_2O(s) + H_2O$$

Although silver fluoride is very soluble in water, AgCl, AgBr, and AgI are quite insoluble, and their solubilities decrease from the chloride to the iodide. The precipitates all darken on exposure to light because of *photochemical decomposition*—decomposition promoted by the absorption of photons of appropriate energy. An example is

Silver iodide is used to "seed" clouds to bring on rain.

$$2AgI \xrightarrow{h\nu} 2Ag + I_2$$

The symbol $h\nu$ over an arrow in a chemical equation means the change is caused by a photon with energy $E = h\nu$.

Silver ion forms many complex ions. In the qualitative analysis for silver, for example, Ag^+ is precipitated as AgCl by the addition of HCl to the "unknown" solution. Ammonia is then added, which causes any AgCl in the precipitate to dissolve. The equilibria involved are

$$AgCl(s) \rightleftharpoons Ag^+(aq) + Cl^-(aq)$$

$$Ag^+(aq) + 2NH_3(aq) \rightleftharpoons Ag(NH_3)_2^+(aq)$$

As NH_3 is added, the second equilibrium is shifted to the right, which decreases the concentration of Ag^+ in the solution. This causes the first equilibrium also to be shifted to the right, and the AgCl dissolves.

In the final step in the test for Ag^+, HNO_3 is added to the mixture thought to contain the $Ag(NH_3)_2^+$. Hydrogen ion from the acid reacts with NH_3 to form NH_4^+. Since this removes molecular NH_3 from the solution, the second reaction above shifts to the left, which increases the Ag^+ concentration, and eventually the Ag^+ and Cl^- recombine to form a precipitate of AgCl.

Chemical Properties of Gold Gold is so unreactive that even concentrated nitric acid fails to attack it. A mixture of concentrated HNO_3 and HCl, called **aqua regia** by the alchemists, will dissolve gold slowly. The chloride ion helps the nitrate ion to oxidize the gold by stabilizing the Au^{3+} ion in the form of a complex ion with the formula $AuCl_4^-$.

Aqua regia is one part concentrated HNO_3 and three parts concentrated HCl, by volume.

SPECIAL TOPIC 3.3 / PHOTOGRAPHY

Compounds of silver are not very stable, and are rather easily decomposed. This is also true of compounds of some other metals—gold, for example—but what makes silver special is that the decomposition of its compounds is promoted by the absorption of light. Despite many attempts, no one has yet found any compounds that quite rival those of silver for light sensitivity, and nearly all the film available today uses silver compounds as the light-sensing medium.

PHOTOGRAPHIC FILM AND PAPER

Photographic film and paper are similar in some respects. Both have a light-sensitive coating called an **emulsion** spread on a supporting base. In film, this base is a flexible clear plastic; in photographic paper the base is either a white paper or, in modern times, a white plastic.

The emulsion consists of gelatin and one or more of the silver halides (AgCl, AgBr, and AgI). The light sensitivity increases from AgCl to AgI, and most negative film uses either AgBr or a mixture of AgBr and AgI. In color film, there are several emulsion layers separated by filter layers to enable the recording of color information. In photographic paper, the emulsion usually contains a mixture of AgBr and AgCl.

Preparation of the photographic emulsion involves precipitation of the silver halides in a metathesis reaction of the type that we discussed in Chapter 4. A solution of $AgNO_3$ is added slowly to a solution that contains KBr (perhaps with a small percentage of KI) and a small amount of gelatin. The net ionic equation is

$$Ag^+(aq) + Br^-(aq) \longrightarrow AgBr(s)$$

The gelatin serves several functions, including keeping the silver halide crystals suspended when the emulsion is spread on the film or paper base.

FORMATION OF THE PHOTOGRAPHIC IMAGE

When you snap a photograph, light falls very briefly on the photographic film. In this brief instant, the silver halide crystals undergo a very subtle change. It is believed that the light absorbed by a crystal causes some electrons to be set free, and these wander through the crystal and become trapped at certain active sites located at the surface. Nearby silver ions then diffuse to these sites and become reduced to silver atoms by the trapped electrons. These silver atoms form the beginnings of the image that will eventually appear on the film. However, at this early stage, no image is yet visible. To the eye, the film would appear unchanged. The still-invisible impression made by the light is called the **latent image,** the term *latent* meaning hidden or dormant.

After you've finished a roll of film, your next step is to bring it to a store to be "developed." The processing of the film involves several chemical reactions, all of which actually take place within the emulsion layer itself. If the film is of the black-and-white variety, the processing is quite simple. The first step is to treat the film with a solution called the "developer," which contains a mild reducing agent. This causes the silver bromide to be reduced to metallic silver, but the reduction occurs most rapidly at those sites where silver atoms of the latent image reside. In a sense, the silver atoms of the latent image serve as nuclei upon which additional silver atoms can be deposited as they are formed in the reduction reaction. As a result, small grains of silver grow only where the film has been exposed to light. The silver particles are so tiny they appear black, and the image is called a *negative* (Figure 1) because those areas that were bright in the original scene show up black on the film.

Photographic developers are normally quite basic. They have to be in order for the special reducing agent to function. Therefore, after the development process is complete and the developer solution has been poured off, the film is treated with a dilute acetic acid solution called a "stop bath." The acid neutralizes the small amount of base in the residual developer ab-

FIGURE 1
A negative and positive print made from it.

sorbed in the emulsion of the film, which halts the development process.

If you were to look at the film after development, you would see the negative image surrounded by milky-white undeveloped silver bromide, which must be removed from the film because it would gradually darken on exposure to light. This is accomplished by a solution called a *fixer,* which contains sodium or ammonium thiosulfate, $Na_2S_2O_3$ or $(NH_4)_2S_2O_3$. The thiosulfate ion reacts with the AgBr and causes it to dissolve by forming a water-soluble complex ion, $Ag(S_2O_3)_2^{3-}$.

$$AgBr(s) + 2S_2O_3^{2-}(aq) \longrightarrow Ag(S_2O_3)_2^{3-}(aq) + Br^-(aq)$$

The fixer, which contains the dissolved AgBr, is then poured off and the film is washed thoroughly with water and dried.

Production of a black-and-white print follows essentially the same sequence as the production of the negative. Light is projected through the negative onto a piece of photographic paper, forming a latent image. The paper is then treated with a developer–stop bath–fixer sequence to produce the final picture. This time a positive image is formed because the paper is not exposed under the dark areas of the negative but is exposed through the clear areas. In this way, dark regions of the negative appear bright in the print, and light areas appear dark, just like in the original scene.

COLOR PHOTOGRAPHS

If you look very closely at a color television screen, you will see that the image is made up of tiny dots of three colors. They are red, green, and blue, and they constitute what are called the **primary colors.** By combining light of just these three colors in the proper proportions we can reproduce any color in the rainbow, and that is exactly what the color TV does. Color film produces its hues by the same principle, with the primary colors being produced by dyes in the film's emulsion.

There are two types of color film that you have probably encountered. Reversal film is the kind used to make color slides. The second type, which is the one we will discuss in some detail below, works much like black-and-white film; the scene is produced on the film as a color negative, which is then printed on special "color paper" to give a positive color print. Both kinds of film are similar in one respect; they contain three emulsion layers. Each is affected by just one of the primary colors. As light from the subject passes through the film, the color image is separated into three images, each being the contribution of just one of the primary colors.

Each light-sensitive layer in a typical color negative film contains a mix of silver halides as the light-sensing medium. Also incorporated into each emulsion layer are portions of dye molecules that are called *dye couplers.* When the film is processed, a special color developer is used. This developer becomes oxidized as it reduces silver ion in places that had been exposed to light. The oxidation products, which are produced in proportion to the amount of silver reduced, combine with the dye couplers in the emulsion layers to produce images made up of dye molecules. Then the silver itself (which is not needed anymore) is removed by oxidizing it to Ag^+ using a mild oxidizing agent, $Fe(CN)_6^{3-}(aq)$, and unexposed silver halides are removed by dissolving them in a thiosulfate fixing bath. This leaves the color negative with its image made up of only dye molecules (Figure 2).

If you examine a color negative (which is somewhat difficult because of its overall orange tint, necessary to improve color quality in the finished print), you will see that it is a "negative" in two ways; light areas in the subject appear dark and the colors of the subject are reversed. Thus, a blue dye covers areas in the subject that are yellow, a red dye covers subject areas that are green, and a green dye covers subject areas that are red. These dyes serve as filters that block light of their opposite color when the negative is printed; blue blocks yellow, red blocks green, and green blocks red. This causes a second reversal in colors when the print is made, and thereby gives the correct colors in the finished photograph (Figure 2).

FIGURE 2

A color photograph and its color negative. In the negative, the colors are reversed; the green leaves appear red, the yellow flower petals appear blue, and the red center of the flower appears green.

$$Au(s) + 6H^+(aq) + 3NO_3^-(aq) + 4Cl^-(aq) \longrightarrow AuCl_4^-(aq) + 3NO_2(g) + 3H_2O$$

Gold is found as the free element in nature.

Although gold forms compounds in both the +1 and +3 oxidation states, none of them is particularly important. In general, gold compounds have little stability and many are easily decomposed by heat.

Zinc, Cadmium, and Mercury

Zn [Ar] $3d^{10}4s^2$
Cd [Kr] $4d^{10}5s^2$
Hg [Xe] $4f^{14}5d^{10}6s^2$

Each of the members of Group IIB has a completed *d* subshell just beneath its outer pair of *s* electrons, and it is only these outer electrons that are involved when zinc reacts. Both electrons are lost, and zinc occurs in compounds only in the +2 oxidation state.

Zinc is found in the ore *zinc blende,* which is zinc sulfide, ZnS. The zinc is recovered by first roasting the ore to give zinc oxide, which is then reduced with carbon. It is a reactive metal, and its surface quickly acquires a film of a basic carbonate, $Zn_2(OH)_2CO_3$. This coating protects the metal below from further oxidation.

The auto industry uses galvanized steel to make "rustproof" automobile bodies.

Zinc metal finds many applications in industry. For example, it is often used to provide a protective coating on steel—a process called **galvanizing.** The zinc is applied either in a thin coat electrolytically or in a much thicker coat by dipping the steel object in molten zinc. The steel is protected in two ways. While the zinc coat is unbroken, of course, the steel is shielded from air and moisture. If the zinc is scratched deeply or worn away, the steel is still kept from corroding by *cathodic protection,* a phenomenon discussed in Chapter 18. Zinc is more easily oxidized than iron, and when the two metals are in contact a galvanic cell is formed in which iron is the cathode and zinc is the anode. Zinc is also used in various alloys. Examples are brass—an alloy of copper and zinc—and bronze—an alloy of copper, tin, and zinc. As noted earlier, pennies today are made of zinc and have just a thin coating of copper. Zinc is also used to make castings and it is important in the manufacture of zinc–carbon dry cells and other batteries.

Recall that oxidation occurs at the anode, so zinc is oxidized.

Zinc is easily oxidized and reacts quickly with nonoxidizing acids to form Zn^{2+} and H_2, and with strong base to give the complex ion $Zn(OH)_4^{2-}$ and H_2.

$$Zn(s) + 2H^+(aq) \longrightarrow Zn^{2+}(aq) + H_2(g)$$

$$Zn(s) + 2OH^-(aq) + 2H_2O \longrightarrow Zn(OH)_4^{2-}(aq) + H_2(g)$$

Zinc hydroxide is also amphoteric. If base is added to a solution containing Zn^{2+} a precipitate of $Zn(OH)_2$ forms, but dissolves on the addition of more strong base.

$$Zn(OH)_2(s) + 2OH^-(aq) \longrightarrow Zn(OH)_4^{2-}(aq)$$

Zinc forms many complex ions similar to its complex with hydroxide ion. For example, if aqueous ammonia is added to $Zn(OH)_2$, the precipitate dissolves with the formation of an ammonia complex.

$$Zn(OH)_2(s) + 4NH_3 \longrightarrow Zn(NH_3)_4^{2+} + 2OH^-(aq)$$

Zinc compounds are used for many purposes. Zinc chloride, which is exceptionally soluble in water (38.1 mol/L at 25 °C), has a wide range of uses that extend from embalming, to fireproofing lumber, to the refining of petroleum. Zinc oxide, a white powder, is used in various creams such as sunscreens and to make quick-setting dental cements. Zinc sulfide is interesting because it can be used to prepare *phosphors*—substances that glow when bathed in ultraviolet light or the high-energy electrons of cathode rays. Such phosphors are used on the inner surfaces of TV picture tubes and the displays of computer monitors, and in devices for detecting atomic radiation.

Cadmium and mercury are both toxic, but zinc is needed by the body.

Cadmium is less abundant than zinc and is usually found as an impurity in zinc ores. The free metal is soft and moderately active. Its chief use is as a protective coating on other metals and for making nickel–cadmium batteries.

The bright red mineral cinnabar, also called vermilion, is composed of mercury(II) sulfide, HgS.

Chemically, cadmium is like zinc in many respects. For example, it dissolves in nonoxidizing acids to form a +2 ion, which is cadmium's only oxidation state in compounds. Unlike zinc, however, cadmium is not affected by base. For this reason, cadmium can be used as a protective coating on metals that must be exposed to an alkaline environment. Cadmium compounds are quite toxic; if absorbed by the body they can cause high blood pressure, heart disease, and even painful death.

Mercury is a metal known since ancient times. Its chief ore is the sulfide, HgS, known as *cinnabar*. Roasting the ore in air gives elemental mercury.

$$HgS(s) + O_2(g) \longrightarrow Hg(l) + SO_2(g)$$

As you know, mercury is a liquid at room temperature. It freezes at −38.9 °C and boils at 357 °C. This large and convenient liquid temperature range accounts for mercury's use as the fluid in thermometers.

A photograph of liquid mercury is on page 60 of the text.

A useful property of mercury is its ability to dissolve many other metals to form solutions called **amalgams.** Dentists use a silver amalgam containing an excess of silver to fill teeth. Gold also forms amalgams easily, and mercury is used to separate gold from its ores. The ore is mixed with mercury, which dissolves the gold. Because mercury is so dense, rocks and other debris float on its surface and can easily be removed. The mercury is then distilled away, leaving the gold behind, and condensed to be used again.

There has been some controversy in recent times about the potential for mercury poisoning from silver fillings in teeth. This issue has not yet been resolved.

Mercury is a much less active metal than either zinc or cadmium. It is unaffected by nonoxidizing acids, but it will dissolve in concentrated nitric acid. The product, $Hg(NO_3)_2$, was once used in the manufacture of felt for hats. Workers often developed severe mercury poisoning, an affliction that leads to nervous disorders, loss of hair and teeth, loss of memory, and even insanity—hence the term "mad as a hatter." Mercury compounds are relatively unstable. The oxide HgO, which forms when mercury is heated at moderate temperatures, decomposes when heated strongly. This thermal instability, in fact, allowed Joseph Priestly to discover oxygen in 1774.

The ease of thermal decomposition of mercury compounds is the reason mercury was discovered so long ago.

In compounds, mercury occurs in two oxidation states, +1 and +2. In the +1 state, corresponding to the loss of only one of its outer *s* electrons, it exists as the Hg_2^{2+} ion (mercurous ion), which contains a mercury–mercury covalent bond.

$$[Hg—Hg]^{2+}$$

Addition of chloride ion to a solution of Hg_2^{2+} gives a white precipitate of Hg_2Cl_2. This reaction is part of the qualitative analytic test for mercury in its +1 oxidation state. Since Ag^+ and Pb^{2+} also give white chloride precipitates, confirmation of mercury(I) is obtained when aqueous ammonia is added, which causes Hg_2Cl_2 to disproportionate.

$$Hg_2Cl_2(s) + 2NH_3(aq) \longrightarrow Hg(l) + Hg(NH_2)Cl(s) + NH_4^+(aq) + Cl^-(aq)$$

The compound $Hg(NH_2)Cl$ is called *mercury amido chloride* and contains mercury in the +2 oxidation state combined with the amide ion, NH_2^-, and Cl^-. The mixture of black finely divided mercury and white $Hg(NH_2)Cl$ usually appears dark gray or black.

Mercury(I) chloride, also known as *calomel,* is very insoluble in water. Its low solubility has permitted therapeutic uses as an antiseptic and as a treatment for syphilis before the discovery of penicillin. Very little mercury is retained by the body because so little Hg_2Cl_2 is able to dissolve. The compound is actually molecular and exists as linear molecules with a mercury–mercury bond.

In the past, Hg_2Cl_2 was used in agriculture to control root maggots on onions and cabbage.

$$:\ddot{\underset{..}{Cl}}—Hg—Hg—\ddot{\underset{..}{Cl}}:$$

Mercury(II) chloride is unusual because even though water soluble it hardly dissociates at all in solution. It exists mostly as molecules of $HgCl_2$ in which covalent bonds exist between Hg and Cl.

This is one exception to the rule that soluble metal salts are fully dissociated into ions in aqueous solution.

$$:\ddot{\underset{..}{Cl}}—Hg—\ddot{\underset{..}{Cl}}:$$

Because it is water soluble, $HgCl_2$ is very poisonous.

Addition of H_2S to a solution containing mercury(II) gives a black precipitate of extremely insoluble HgS. This is an interesting compound because when it is heated its crystal structure changes and it becomes a brilliant red substance called *vermilion.*

SUMMARY

Metals Metals form a metallic lattice in which valence electrons are shared among all the metal atoms in the crystal. Metals tend to form ionic compounds with nonmetals. Within the periodic table, **metallic character** increases from right to left in a period and from top to bottom in a group.

Metallurgy Some unreactive metals, such as gold, are found free in nature, but most are combined with other elements. Sodium and magnesium are extracted from seawater. **Manganese nodules** on the ocean floor are a potential source of manganese and iron. Usually, when an **ore** is dug from the ground, the metal-bearing component must be enriched by a pretreatment step that removes much of the unwanted **gangue. Flotation** is often used with lead and copper sulfide ores. Sulfide ores are usually **roasted** to convert them to oxides, which are more easily reduced. Aluminum's amphoteric character is exploited in purifying bauxite.

Metals are obtained from their compounds by reduction. Active metals such as sodium, magnesium, and aluminum must generally be prepared by electrolysis. Carbon, in the form of **coke,** is a common chemical reducing agent because it is plentiful and inexpensive. Metallic iron forms in a **blast furnace** where a charge of iron ore, limestone, and coke reacts in a stream of heated air. Molten iron and **slag** flow to the bottom of the furnace and are periodically tapped. **Steel** is made from this impure iron by removing impurities and adding special metals. The **open hearth furnace** has been replaced by the **basic oxygen process** as the most frequently used method for making steel.

The Alkali Metals Sodium and potassium are the most common of the alkali metals, all of which are soft, low melting, low density, and very reactive. Lithium has an exceptionally negative reduction potential because of its small size, which causes it to have a very large hydration energy. Lithium and sodium are prepared by electrolysis; the heavier alkali metals are displaced from their chlorides by sodium vapor.

All the alkali metals react with water to form H_2 and the metal hydroxide. Only lithium reacts with N_2 to form lithium nitride, and only lithium forms the normal oxide on reaction with O_2. Sodium forms the peroxide, Na_2O_2, and the rest form superoxides, MO_2.

The most important compounds are the chlorides, hydroxides, and carbonates. The most plentiful alkali metal compound is NaCl, so sodium compounds are much less expensive than potassium compounds. Electrolysis of aqueous NaCl gives NaOH. KOH is made the same way. Sodium carbonate is made from *trona* ($Na_2CO_3 \cdot NaHCO_3 \cdot 2H_2O$) and by the **Solvay** process. $NaHCO_3$ (baking soda) decomposes on heating to give Na_2CO_3.

Sodium salts of strong acids tend to be more soluble in water than the potassium salts, whereas potassium salts of weak acids tend to be more water soluble than the sodium salts. Sodium salts impart a bright yellow color to a flame, lithium salts give a brilliant red, and potassium salts give a violet color.

Alkaline Earth Metals These Group IIA metals are harder, higher melting, higher density, and less reactive than their neighbors in Group IA. Only magnesium and beryllium are made in commercial quantities, because they don't react with air and moisture. Magnesium is obtained from **dolomite** and seawater; the metal is formed in the electrolysis of $MgCl_2$. Beryllium is obtained from **beryl**, $[Be_3Al_2(SiO_3)_6]$, which forms the gemstones emerald and aquamarine. Electrolysis of $BeCl_2$ with NaCl as an electrolyte gives Be. **Limestone** ($CaCO_3$), dolomite ($CaCO_3 \cdot MgCO_3$), and **gypsum** ($CaSO_4 \cdot 2H_2O$) are rich sources of calcium.

Beryllium does not react with water because of its tough insoluble oxide coating. Magnesium reacts with hot water and steam, but not with cold water. Its surface is also protected by its oxide. Calcium, strontium, and barium react with water to give H_2 and the metal hydroxide. All beryllium compounds are covalent because of the small size of the Be^{2+} ion. Beryllium and its oxide are amphoteric. Nitrides with the general formula M_3N_2 are formed by Mg, Ca, Sr, and Ba by direct reaction of the metals with N_2.

All the chlorides are water soluble and all are ionic except for $BeCl_2$, which is covalent and exists in the solid state as **polymeric** $(BeCl_2)_x$. $CaCl_2$ is **deliquescent.** The oxides are usually made by decomposition of the carbonates. Calcium oxide is called **lime** and reacts with water to form the hydroxide **(slaked lime).** Magnesium hydroxide is in **milk of magnesia.** The hydroxides increase in solubility from $Mg(OH)_2$ to $Ba(OH)_2$. The sulfates decrease in solubility from $MgSO_4$ to $BaSO_4$. Gypsum can be partially dehydrated to give **plaster of Paris.** Barium sulfate is used to obtain medical X-ray photographs of the intestinal tract.

Calcium salts give a red-orange color to a flame, strontium salts give a crimson color, and barium salts give a yellow-green color.

Aluminum Aluminum is a very common element in the Earth's crust but can only be obtained easily from a few minerals. Its chief ore is **bauxite,** Al_2O_3. Aluminum is easily oxidized to the +3 state, and in air it forms an oxide coating that protects the metal from further attack. In the **thermite reaction,** the large heat of formation of Al_2O_3 is used to reduce iron oxide to give elemental iron in a molten state. Aluminum is amphoteric. In acidic solution or neutral solution it exists as the complex ion $Al(H_2O)_6^{3+}$. This ion is itself a weak acid. When a solution that contains $Al(H_2O)_6^{3+}$ is made basic, a gelatinous precipitate with the formula $Al(H_2O)_3(OH)_3$ is formed that can dissolve in either acid or base.

Aluminum oxide occurs in two crystalline forms, γ-Al_2O_3, which is soluble in acids and bases, and α-Al_2O_3 **(corundum),** which is quite inert. Gems such as ruby and sapphire are composed of α-Al_2O_3. Aluminum sulfate is made in large quantities by dissolving bauxite in sulfuric acid and forms double salts called **alums,** for example, $KAl(SO_4)_2 \cdot 12H_2O$.

Aluminum is similar in many ways to beryllium. Its anhydrous chloride and bromide are dimeric with the general formula Al_2X_6.

Post-transition Metals These elements are characterized by two oxidation states. The lower one comes from losing the outer *p* electron(s) and the higher comes from the further loss of the outer pair of *s* electrons. Going down a group, the lower oxidation state becomes progressively more stable relative to the higher one. In Group IIIA, the oxidation states observed are +1 and +3. In Group IVA they are +2 and +4, and in Group VA, bismuth has +3 and +5 oxidation states.

Tin is made from its oxide SnO_2 by reduction with carbon and is purified electrolytically. It forms three allotropes. **White tin** is metallic; **gray tin** has the diamond crystal structure and has nonmetallic properties. When heated, the metallic allotrope changes to **brittle tin.** Tin reacts with oxygen and oxidizing acids such as HNO_3 to form SnO_2. Two kinds of halides are formed, SnX_2 and SnX_4. The tetrahalides such as $SnCl_4$ are covalent. For tin, the +4 oxidation state is easily formed from the +2 state.

Lead is made from its sulfide PbS which is first roasted to give the oxide PbO and then reduced with carbon. It is also purified electrolytically. Lead occurs only in a metallic form. Three oxides are known: PbO **(litharge),** PbO_2, and a mixed oxide Pb_3O_4 (called **red lead**). The +2 state is much more stable than the +4 state for lead, and lead(IV) compounds are very strong oxidizing agents. The halides PbX_2 can be made and are relatively insoluble in water. $PbBr_4$ and PbI_4 are very unstable because lead(IV) is so strong an oxidizing agent; $PbCl_4$ decomposes slowly.

Bismuth is obtained from its oxide or sulfide. The sulfide is roasted to give the oxide, which is then reduced with carbon. Bismuth is also a by-product of the production of lead. The metal is used to make low-melting alloys such as **Wood's metal.** Bismuth reacts with oxygen and chlorine to form Bi_2O_3 and $BiCl_3$, respectively. The chloride, $BiCl_3$, hydrolyzes in water to give a precipitate of bismuthyl chloride, BiOCl. The +5 state can only be made under severely oxidizing conditions, and bismuth(V) is a very powerful oxidizing agent. Sodium bismuthate, $NaBiO_3$, is used in the qualitative test for manganese where it oxidizes Mn^{2+} to MnO_4^-.

Transition Elements: General Characteristics and Periodic Trends The elements in the center of the periodic table are the transition elements. They have partially filled *d* subshells (except for Zn, Cd, and Hg); they are generally hard and have high melting points; they exhibit multiple oxidation states; their ions form many complex ions; and many of their compounds are colored. The sizes of the transition metal atoms decrease gradually from left to right across the table. The transition elements in period 6 are nearly the same size as those above them in period 5 because of the **lanthanide contraction.** As a result, the transition elements in period 6 that occur after the lanthanide series are very dense and unreactive because of the large effective nuclear charge felt by their outer electrons.

The **ferromagnetic** elements are iron, cobalt, and nickel. In their crystals, the paramagnetic atoms have their magnetic poles aligned within regions called **domains.** In an unmagnetized state, the domains point in random directions, but when "permanently" magnetized the domains are aligned.

Properties of Important Transition Metals Many common substances are composed of transition metal compounds or are the transition metals themselves.

Titanium This metal is lightweight and resists corrosion. Its chloride, $TiCl_4$, is molecular and reacts readily with water to give TiO_2. Titanium dioxide is an important paint pigment.

Vanadium Alloy steels contain vanadium to give them ductility and shock resistance. Vanadium pentaoxide, V_2O_5, is the catalyst in the contact process for the manufacture of sulfuric acid. Solutions of V_2O_5 are acidic and contain the VO_2^+ ion, which can be viewed as forming by extensive hydrolysis of a V^{5+} cation.

Chromium Chromium has many properties that make it useful as a protective coating over other metals. It is used in *stainless steel* alloys and to make *nichrome* heating elements. The metal is moderately active and dissolves in nonoxidizing acids, yielding Cr^{2+} and H_2 in the absence of oxygen. All of chromium's compounds are colored. Its principal oxidation states are +2, +3, and +6. Chromium(II) is easily oxidized to Cr^{3+}, which is amphoteric. $Cr(H_2O)_6^{3+}$ precipitates a gelatinous hydroxide in base, which redissolves in excess base. Chromium(III) forms many complexes; the $Cr(H_2O)_6^{3+}$ ion is blue-purple and is found often in chromium(III) salts. CrO_3 is the acid anhydride of chromic acid, H_2CrO_4, which exists at very low pH. At higher pH there is an equilibrium between $Cr_2O_7^{2-}$ and CrO_4^{2-}. The oxide Cr_2O_3 is green and is used extensively as a pigment.

Manganese Manganese corrodes in moist air and dissolves readily in nonoxidizing acids to give H_2 and Mn^{2+}. It is used mostly in alloys. The most important oxidation states of manganese are +2, +3, +4, +6, and +7. The +2 state is the most stable because the ion has a half-filled *d* subshell. When precipitated as $Mn(OH)_2$, oxidation to $MnO(OH)$ occurs readily. Mn^{3+} is unstable in water, but can be stabilized by the formation of complex ions. The least stable oxidation state is +7 in the MnO_4^- ion. In acidic solutions, this permanganate ion is reduced to Mn^{2+} and in basic or neutral solution it is reduced to MnO_2. MnO_2 is a nonstoichiometric compound. Oxidation of MnO_2 by air or KNO_3 in molten KOH gives the MnO_4^{2-} ion. The MnO_4^{2-} ion is stable only in basic solution; it disproportionates in acidic solution.

Iron Iron is a common metal and the one most familiar to people. It is moderately active and dissolves in nonoxidizing acids to give H_2 and Fe^{2+}. Its chief ores contain hematite, $Fe_2O_3 \cdot xH_2O$, and magnetite, Fe_3O_4, the oxide in the more recently used taconite. The principal oxidation states of iron are +2 and +3. Oxides include FeO (difficult to prepare), Fe_2O_3, and Fe_3O_4. The ion $Fe(H_2O)_6^{3+}$ is pale green and occurs in various iron(II) salts. Solutions of $Fe(H_2O)_6^{3+}$ are acidic and slightly yellow because of hydrolysis. Addition of base precipitates a gelatinous hydroxide. Iron forms many complex ions. Examples given are $Fe(SCN)_6^{3-}$, $Fe(CN)_6^{4-}$, and $Fe(CN)_6^{3-}$. The reaction of Fe^{2+} with $Fe(CN)_6^{3-}$ gives a blue precipitate of *Prussian blue,* a reaction that is used to make blueprints.

Cobalt Cobalt is less reactive than iron but does dissolve slowly in nonoxidizing acids. Its chief use is in alloys. Two oxidation states are important, +2 and +3. In water, Co^{2+} exists as the pink ion $Co(H_2O)_6^{2+}$. Heating pink $[Co(H_2O)_6]Cl_2$ changes the color to blue as dehydration gives $[Co(H_2O)_4]Cl_2$. Cobalt(III) ion oxidizes water, but forms many stable complex ions.

Nickel The metal is resistant to corrosion and forms alloys with iron that are impact resistant. Purification of nickel by the **Mond process** relies on the ease of formation of nickel carbonyl, $Ni(CO)_4$. The metal dissolves in nonoxidizing acids with the evolution of H_2 and the formation of Ni^{2+}. In solution the $Ni(H_2O)_6^{2+}$ ion is green. The only particularly stable oxidation state of nickel is +2. The compound NiO_2 is used as the cathode in nickel–cadmium batteries.

Copper, Silver, and Gold These are known as the **coinage metals** and they are less reactive than the other metals discussed. They do not dissolve in nonoxidizing acids, but copper and silver do dissolve in nitric acid. Gold dissolves in **aqua regia.** Copper has a high electrical and thermal conductivity and is used to make electrical wire and pipe. Silver is an even better conductor, but it's too expensive to use in these applications. Gold is used mainly as a decorative metal and in specialized electrical applications. Gold leaf is extremely thin gold that's used for decorative purposes.

Copper has two oxidation states, +1 and +2. Extended exposure of the metal to air gives it a green coating of $Cu_2(OH)_2CO_3$. Heating the metal in air yields the black oxide CuO. The Cu^+ ion disproportionates in water, but can be isolated in solids such as CuCl. The most stable state is +2, which forms many complex ions. In water, copper(II) ion exists as $Cu(H_2O)_4^{2+}$.

Silver tarnishes in air and forms a thin film of Ag_2S. The only important oxidation state is +1. Addition of base to Ag^+ gives Ag_2O. AgF is soluble, but the other halides are insoluble and light sensitive. The Ag^+ ion forms many complex ions in which it is usually bonded to two ligands.

Gold is found free in nature. It has oxidation states of +1 and +3, but none of its compounds is particularly important. Gold compounds are easily decomposed by heat.

Zinc, Cadmium, and Mercury Zinc occurs in the ore zinc blende, ZnS, which is roasted to the oxide and reduced with carbon. Zn is very reactive and dissolves in both acid and base to give H_2. By reaction with air and moisture, the metal forms a protective coating of $Zn_2(OH)_2CO_3$. One of the chief uses of zinc is in **galvanizing** steel, where it protects the steel against corrosion by cathodic protection. It is also used in alloys such as brass and bronze, and in batteries. Zinc and its hydroxide dissolve in acids and bases. The only oxidation state is +2, and the Zn^{2+} ion forms many complexes.

Cadmium is less common than zinc and occurs as an impurity in zinc ores. It is moderately active and dissolves in nonoxidizing acids, but it does not dissolve in base. Cadmium compounds are very toxic.

Mercury is a liquid over a broad range of temperatures and is used in thermometers. It also forms **amalgams,** which are solutions of metals in mercury. Mercury does not react with nonoxidizing acids but does dissolve in concentrated HNO_3 to give $Hg(NO_3)_2$. There are two oxidation states, +1 (Hg_2^{2+} ion) and +2 (Hg^{2+} ion). The Hg_2^{2+} ion has a mercury–mercury covalent bond. Hg_2Cl_2 is insoluble in water and disproportionates when aqueous ammonia is added; the products are Hg and $Hg(NH_2)Cl$. The very poisonous $HgCl_2$ is water soluble, but only a very weak electrolyte. Both Hg_2Cl_2 and $HgCl_2$ are linear covalent molecules.

REVIEW EXERCISES

Answers to questions whose numbers are printed in color are given in Appendix A. More challenging questions are marked with asterisks.

Metallic Character and the Periodic Table

3.1 What is a metallic lattice? Why do metals form this kind of solid?

3.2 How does the metallic character of the elements vary (a) going from left to right across a period and (b) going from top to bottom in a group?

3.3 Choose the more metallic element in each pair: (a) Ga or Tl, (b) Ba or Tl, (c) Mg or K.

3.4 Which compound is most ionic, $MgCl_2$, $AlCl_3$, or RbCl?

3.5 Which oxide is more basic, BeO or MgO?

3.6 Which oxide is more acidic, MgO or Al_2O_3?

Metallurgy

3.7 Use your own words to define *metallurgy.*

3.8 What is an *ore?* What distinguishes an ore from some other potential source of a metal?

3.9 Many rocks are composed of minerals called aluminosilicates. One such mineral is called *orthoclase* and has the formula $KAlSi_3O_8$. Despite their high abundance, aluminosilicates are not considered aluminum ores. What is the probable reason for this?

3.10 Why are sodium and magnesium extracted from seawater?

3.11 Write chemical equations for the reactions that are used to obtain metallic magnesium from seawater.

3.12 What is *lime*? How is it made from $CaCO_3$?

3.13 What are *manganese nodules?* Why are they a potential source of metals? Why aren't they being mined now?

3.14 What is *gangue?*

3.15 Why can gold be separated from impurities of rock and sand by panning?

3.16 Describe the *flotation process.*

3.17 Write chemical equations for the reactions that occur when Cu_2S and PbS are roasted in air. Write a chemical equation to show how SO_2 from the roasting can be kept from being released to the environment. Why might a sulfuric acid plant be located near a plant that roasts sulfide ores?

3.18 Write chemical equations that show how bauxite is purified.

3.19 Why is reduction, rather than oxidation, necessary to extract metals from their compounds?

3.20 Sodium, magnesium, and aluminum are produced by electrolysis instead of by reduction with chemical reducing agents. Why?

3.21 Why is carbon such a useful industrial reducing agent?

3.22 What is *coke*? How is it made?

3.23 Write chemical equations for the reduction of PbO and CuO with carbon.

3.24 Copper(I) sulfide can be converted to metallic copper without adding a reducing agent. Explain this using appropriate chemical equations.

3.25 Why is a blast furnace called a *blast* furnace?

3.26 What is the composition of the charge that's added to a blast furnace?

3.27 Describe the chemical reactions involved in the reduction of Fe_2O_3 that take place in a blast furnace. What is the active reducing agent in the blast furnace?

3.28 What is slag? Write a chemical equation for its formation in a blast furnace. What are some of its uses?

3.29 What does *refining* mean in metallurgy?

3.30 What is the difference between pig iron and steel?

3.31 What is a *Bessemer converter?* Why was it replaced by the open hearth furnace?

3.32 Describe the operation of an open hearth furnace. What reactions take place during the production of steel?

3.33 Describe the basic oxygen process. Why has it replaced the open hearth furnace as the chief steel-making method?

The Alkali Metals

3.34 Write the electron configurations of (a) Li, (b) K, and (c) Fr.

3.35 What are the sources of the alkali metals? Why aren't these elements found free in nature?

3.36 What is *saltpeter?* What is *Chile saltpeter?*

3.37 Why is francium such a rare metal?

3.38 Which of the alkali metals are most important biologically? Which is most important in plants?

3.39 Why are the alkali metals soft and low melting?

3.40 Which alkali metal has been used as a coolant in a nuclear reactor? Why?

3.41 Explain why lithium has an unexpectedly large negative standard reduction potential. In the absence of water, which is expected to be the better reducing agent, lithium or sodium? Why?

3.42 How is metallic sodium prepared? Why isn't metallic potassium prepared in the same way? Why can molten KCl be reduced by sodium vapor?

3.43 Complete and balance the following chemical equations.

(a) $Li + H_2O \rightarrow$
(b) $K + H_2O \rightarrow$
(c) $Rb + H_2O \rightarrow$

3.44 Complete and balance the following chemical equations.

(a) $Li + O_2 \rightarrow$
(b) $Na + O_2 \rightarrow$
(c) $Rb + O_2 \rightarrow$

3.45 Complete and balance the following chemical equations.

(a) $Na_2O_2 + H_2O \rightarrow$
(b) $Na_2O + H_2O \rightarrow$
(c) $KO_2 + H_2O \rightarrow$
(d) $NaOH + CO_2 \rightarrow$

3.46 Which alkali metal reacts with N_2? Write the chemical equation for the reaction.

3.47 What are the formulas for (a) caustic soda and (b) lye?

3.48 What reactions take place at the anode and cathode in the mercury cell during the electrolysis of aqueous NaCl? Write a chemical equation that shows how NaOH is formed in this apparatus. (If necessary, refer back to Chapter 18.)

3.49 What is the principal use of Na_2CO_3? Give the name and formula for the ore used to obtain Na_2CO_3.

3.50 Use chemical equations to describe how Na_2CO_3 is made by the Solvay process.

3.51 Write a chemical equation that shows how $NaHCO_3$ functions during the baking of bread.

3.52 Write chemical equations that show how $NaHCO_3$ functions as a buffer.

3.53 Use a chemical equation to show why solutions of Na_2CO_3 are basic. Calculate the pH of a 1.0 *M* solution of Na_2CO_3.

3.54 Which salt would you expect to be more water-soluble in each of the following pairs? (a) $NaClO_4$ or $KClO_4$, (b) $NaC_2H_3O_2$ or $KC_2H_3O_2$, (c) $KC_2H_3O_2$ or $KClO_4$.

3.55 What color flame is produced by (a) Na^+, (b) K^+, and (c) Li^+?

3.56 In April 1983, 20,000 gallons of nitric acid spilled from a ruptured railroad car at a rail siding near downtown Denver, Colorado. Giant snowblowers were used to blow soda ash on the spill to neutralize the acid. Write a chemical equation for this neutralization reaction.

***3.57** Given that concentrated nitric acid has a concentration of 16 *M*, how many pounds of soda ash were required to completely react with all the HNO_3 spilled by the tank car described in Review Exercise 3.56?

Alkaline Earth Metals

3.58 Write the electron configurations of (a) Mg and (b) Sr.

3.59 What are the electron configurations of the ions Mg^{2+} and Ca^{2+}?

3.60 Are the alkaline earth metals ever found free in nature? Explain your answer.

3.61 Give the formulas for (a) limestone, (b) dolomite, (c) gypsum, and (d) beryl.

3.62 What is the source of radium?

3.63 Which gems are composed of beryl?

3.64 Why are the alkaline earth metals harder and higher melting than the alkali metals alongside? How does the ease of oxidation of the Group IIA metals compare with that of the Group IA metals?

3.65 How are the alkaline earth metals prepared?

3.66 Which alkaline earth metals react with cold water? Write the chemical equation for the reaction of one of them with water.

3.67 Which of the alkaline earth metals react only with hot water? Which do not react with water at all?

3.68 Which are the only alkaline earth metals to be prepared in significant quantities? Why these and not the others?

3.69 How is beryllium metal prepared?

3.70 Why is beryllium added to copper or bronze? What is a typical use for one of these alloys?

3.71 Show by means of chemical equations how magnesium is extracted from dolomite.

3.72 Which alkaline earth metal is used in flashbulbs and flares? Why? Explain using a chemical equation.

3.73 How are the properties of magnesium and lithium alike? What is the apparent reason? How do the properties of magnesium and lithium differ?

3.74 Why is $BeCl_2$ covalent, but $MgCl_2$ is ionic? How does the tendency to form covalent bonds vary among the Group IIA elements?

3.75 Which of the alkaline earth metals react with elemental nitrogen? Write an equation for one of these reactions.

3.76 Describe the structure of $BeCl_2$ in the solid state.

3.77 What compound of a Group IIA element is found in milk of magnesia? Write a chemical equation to show how this compound neutralizes stomach acid, HCl.

3.78 How do the solubilities of the Group IIA sulfates and hydroxides vary going from top to bottom in the group?

3.79 Define *deliquescence.* Write the chemical formula for an alkaline earth metal compound that is *deliquescent.*

3.80 Write chemical equations for two reactions that give MgO as a product.

3.81 Which compound of the alkaline earth metals is used to make refractory bricks? Why?

3.82 How can $Ca(OH)_2$ and $Mg(OH)_2$ be prepared?

3.83 Give uses for the following:

(a) $CaSO_4 \cdot 2H_2O$
(b) $CaCO_3$
(c) $BaSO_4$
(d) $MgSO_4 \cdot 7H_2O$
(e) $CaCl_2$
(f) $Mg(OH)_2$
(g) Be
(h) Mg

3.84 How is plaster of Paris made? Write its formula and write a chemical equation for the reaction of plaster of Paris with H_2O.

3.85 What color flame is produced by (a) Ca^{2+}, (b) Sr^{2+}, and (c) Ba^{2+}?

Aluminum

3.86 Write the electron configurations for (a) Al and (b) Al^{3+}.

3.87 How does the abundance of aluminum in the Earth's crust compare with (a) that of other elements and (b) that of other metals?

3.88 How is aluminum extracted from its ores?

3.89 What is the formula of the mineral *cryolite*? What ions are present in it?

3.90 Aluminum is a very reactive metal. Why doesn't it experience rapid and extensive corrosion in air?

3.91 What is anodized aluminum?

3.92 Write the chemical equation for the thermite reaction.

3.93 Why is aluminum used as a fuel in the rocket propellant in the booster rockets of the space shuttle?

3.94 Write net ionic equations for the reactions of aluminum with aqueous solutions of (a) HCl and (b) NaOH.

3.95 What is the formula of aluminum-containing species that exist in acidic or neutral solutions? Why do many aluminum salts crystallize from aqueous solutions as hydrates?

3.96 Why is the $Al(H_2O)_6{}^{3+}$ ion acidic? Write chemical equations that show what happens to the $Al(H_2O)_6{}^{3+}$ ion when a base is added gradually to a solution that contains it.

3.97 Write chemical equations that show what happens when aluminum hydroxide is dissolved in (a) a base and (b) an acid.

3.98 How is aluminum hydroxide used in water purification? How is it formed?

3.99 In a strongly basic solution, aluminum exists as an ion that can be written as $AlO_2{}^-$. What is a better way of writing the formula for this aluminum-containing ion?

3.100 What differences are there in the properties of γ-Al_2O_3 and α-Al_2O_3?

3.101 What ions are responsible for the colors of ruby and sapphire?

3.102 What is an *alum*? Why are alums called double salts? Write the chemical formulas for sodium alum and potassium alum. What are some of their uses?

3.103 Sketch the structure of the dimer of Al_2Cl_6. How are the chlorines arranged geometrically around the aluminum atoms?

3.104 How is the structure of Al_2Cl_6 similar to the structure of $BeCl_2$ in the solid state? Specify one other way the chemistries of beryllium and aluminum are similar.

Other Metals in Groups IIIA, IVA, and VA

3.105 Which are the post-transition metals? Write their electron configurations.

3.106 What oxidation states are observed for (a) Tl, (b) Sn, (c) Pb, and (d) Bi? Account for them in terms of the electron configurations of the elements.

3.107 How do the relative stabilities of the oxidation states of Ga, In, and Tl vary? How is this explained?

3.108 In general terms, why are the reactivities of the post-transition metals lower than the reactivities of the metals in Groups IA and IIA?

3.109 Write chemical equations for the preparation of tin, lead, and bismuth from their ores.

3.110 What are the allotropes of tin? What are their properties?

3.111 Which allotrope of tin is used to coat a "tin can"?

3.112 What property of bismuth makes it useful for making castings?

3.113 Why is $SnCl_4$ covalent, but $SnCl_2$ is ionic?

3.114 Write the equation for the reaction of tin with hydrochloric acid.

3.115 Write chemical equations for the reactions of tin and lead with HNO_3.

3.116 Which is a better oxidizing agent, SnO_2 or PbO_2?

3.117 Write the chemical formulas for the stannite ion and the plumbite ion. Which is the better reducing agent?

3.118 Write net ionic equations for the reactions of a strong base with (a) Sn, (b) SnO, (c) Pb, and (d) PbO.

3.119 What is the formula for tetraethyllead? What is it used for?

3.120 What are the oxides of lead? Give one use of each.

3.121 The reaction of plumbite ion with hypochlorite ion in basic solution gives solid PbO_2 and chloride ion. Write a balanced net ionic equation for the reaction.

3.122 When concentrated HCl is added to solid PbO_2, a reaction occurs in which Cl_2 gas is evolved and $PbCl_2$ is formed. Write a balanced net ionic equation for the reaction.

3.123 What is the average oxidation number of lead in Pb_3O_4? What mixture of oxides can be considered to make up this oxide?

3.124 What is the formula for bis(tributyltin) oxide? Why isn't it used anymore in paints intended for marine environments?

3.125 Write chemical equations for the reactions of bismuth with (a) O_2 and (b) Cl_2.

3.126 Use the data in Table 17.2 to calculate the molar solubilities in water of (a) $PbCl_2$, (b) $PbBr_2$, and (c) PbI_2.

3.127 Why doesn't $BiCl_5$ exist?

3.128 Why are $PbBr_4$ and PbI_4 so unstable? Write an equation for the decomposition of $PbBr_4$.

3.129 What is the formula for "white lead"? Why does it gradually darken in air when it is used as a paint pigment?

3.130 What lead-containing pigment used by artists is yellow?

3.131 Write the chemical equation for the hydrolysis of $BiCl_3$. Explain why adding concentrated HCl to a solution that contains this hydrolysis reaction causes BiOCl to dissolve.

3.132 What chemical reaction is involved in the qualitative test for Mn^{2+}? Describe what is observed during the test.

General Properties of Transition Metals

3.133 Make a sketch of the periodic table. Mark off those regions where you would find (a) the transition elements, (b) the inner transition elements, (c) the lanthanides, (d) the actinides, and (e) the representative elements.

3.134 Give three properties of the transition elements that distinguish them from the representative elements.

3.135 Why is a +2 oxidation state common among the transition elements?

3.136 What is a complex ion? What would be the formula and the charge on a complex ion formed from Cu^{2+} and four OH^- ions?

3.137 Why are many of the transition elements able to exist in more than one oxidation state?

3.138 How do the sizes of the transition metal atoms vary from left to right across a period?

3.139 What is the *lanthanide contraction?* How does it affect the properties of the transition elements in period 6?

3.140 What is the origin of the lanthanide contraction?

3.141 What similarities exist between elements in the A and B groups in the periodic table? Give three examples.

3.142 The density of niobium (Nb) is 8.57 g/cm^3. Both Nb and Ta atoms are the same size. Estimate the density of tantalum.

3.143 In the periodic table, what is a *triad?* Where do we find triads?

3.144 Which elements in their pure states are ferromagnetic?

3.145 Why do paramagnetic substances feel such a weak attraction to a magnet compared with ferromagnetic substances?

3.146 Why does iron lose its ferromagnetic properties and become paramagnetic when melted?

3.147 If an iron bar is lined up with its long axis pointing north–south and then given repeated sharp blows with a hammer, it becomes a weak permanent magnet. Explain what happens within the iron bar when this occurs.

Chemistry of Some Important Transition Elements: Titanium, Vanadium, and Chromium

3.148 Why is titanium a commercially useful metal? Why can't its oxide be reduced with carbon by ordinary metallurgical methods? How is metallic titanium made?

3.149 What titanium compound was used by the U.S. Navy to make smoke screens? Write the chemical equation for the reaction that produces the smoke.

3.150 Explain why $TiCl_4$ is covalent.

3.151 What is the formula for the most commercially important compound of titanium? What is its principal use?

3.152 What are two principal uses of elemental vanadium?

3.153 What is V_2O_5 used for?

3.154 In Section 3.5 we described the hydrolysis of $BiCl_3$ to give BiOCl. Explain how the BiO^+ ion can be accounted for in terms of the hydrolysis of the Bi^{3+} ion.

3.155 What vanadium-containing species is formed in aqueous solutions of V_2O_5?

3.156 Why is elemental chromium used to plate automobile parts?

3.157 What is *nichrome?* What is it used for?

3.158 What are the common oxidation states of chromium? Write their electron configurations.

3.159 What is chrome alum? How can it be made? What color is it?

3.160 Write a balanced net ionic equation for the oxidation of Cr^{2+} by O_2 in an acidic solution.

3.161 Which is a better reducing agent, Cr^{2+} or Cr^{3+}?

3.162 Write a chemical equation that explains why solutions of Cr^{3+} are slightly acidic.

3.163 Write chemical equations to show what happens

when concentrated NaOH solution is gradually added to a solution containing chromium(III) ion. Identify the colors of the different species.

3.164 What relationships exist among CrO_3, H_2CrO_4, CrO_4^{2-}, and $Cr_2O_7^{2-}$?

3.165 What are the colors of the CrO_4^{2-} and $Cr_2O_7^{2-}$ ions?

3.166 Write the equilibria for the ionization of chromic acid.

3.167 The oxide Cr_2O_3 is amphoteric, meaning it is both acidic and basic, but CrO_3 is only acidic. Why is this? Explain, using appropriate Lewis structures, how CrO_4^{2-} is changed to $Cr_2O_7^{2-}$ as the pH is lowered.

3.168 What color is Cr_2O_3? What is it used for?

3.169 What are the names of (a) H_2CrO_4, (b) $Cr_2O_7^{2-}$, (c) CrO_4^{2-}, and (d) Cr_2O_3?

3.170 Write the formulas of two yellow chromium-containing pigments.

Chemistry of Some Important Transition Elements: Manganese, Iron, Cobalt, and Nickel

3.171 What are the common oxidation states of manganese? Which is the most stable? Which is most easily reduced to a lower oxidation state?

3.172 How does the reactivity of manganese compare with that of chromium?

3.173 Write molecular equations for the reactions of (a) manganese with hydrochloric acid and (b) manganese(II) chloride with aqueous NaOH.

3.174 What product is formed when freshly precipitated $Mn(OH)_2$ is exposed to air?

3.175 What are the colors of (a) $Mn^{2+}(aq)$, (b) MnO_4^-, and (c) MnO_4^{2-}?

3.176 What are the names of (a) MnO_4^-, (b) MnO_4^{2-}, and (c) MnO_2?

3.177 Under what conditions is the manganate ion stable?

3.178 What properties of MnO_4^- make it particularly useful in redox titrations?

3.179 What is a *nonstoichiometric compound?* Give an example of one.

3.180 Write half-reactions for the reduction of MnO_4^- in (a) an acidic solution and (b) a basic solution.

3.181 What is a *disproportionation reaction?* Write a chemical equation for the disproportionation of MnO_4^{2-} in acidic solution.

3.182 Which compound of manganese is used in the common dry cell? What is its function?

3.183 What compounds are found in ores of iron?

3.184 What happens when iron is treated with (a) concentrated hydrochloric acid and (b) concentrated nitric acid?

3.185 What are the common oxidation states of iron?

3.186 What are the formulas of the oxides of iron?

3.187 Write a net ionic equation for the reaction of iron with a solution of HCl.

3.188 Write a net ionic equation for the reaction that occurs when base is added to aqueous solutions of (a) ferrous ion and (b) ferric ion.

3.189 What happens to freshly precipitated $Fe(OH)_2$ if it is exposed to air?

3.190 Write formulas for (a) rust, (b) the ion that gives a green color to an aqueous solution of iron(II) salts, (c) Prussian blue, and (d) the gelatinous precipitate that forms when NaOH is added to a solution of iron(III) sulfate.

3.191 Why are solutions of Fe^{3+} slightly acidic?

3.192 Why are solutions of iron(III) salts often slightly yellow?

3.193 In what significant way do $Fe(OH)_3(H_2O)_3$ and $Cr(OH)_3(H_2O)_3$ differ chemically?

3.194 Explain why a solution of Fe^{3+} is more acidic than a solution of Fe^{2+}.

3.195 What are the common names for $K_4Fe(CN)_6$ and $K_3Fe(CN)_6$? What are their IUPAC names?

3.196 What reactions are involved in the rusting of iron?

3.197 What chemicals are involved in making a blueprint? Describe the chemical reactions that are involved.

3.198 On the basis of what you have learned about the chemistry of iron in this chapter, describe how you could test a solution for the presence of cyanide ion, CN^-.

3.199 What are the properties and uses of cobalt?

3.200 What is *stellite?* What is *alnico?*

3.201 What color is cobalt(II) ion in aqueous solution? What ion is responsible for the color?

3.202 Write a chemical equation that shows what happens when $CoCl_2 \cdot 6H_2O$ is heated. What color change takes place?

3.203 Write a balanced net ionic equation for the oxidation of water by Co^{3+} ion in acidic solution.

3.204 Why is nickel an important metal? What metals are in the "nickel" used to make coins? What is monel?

3.205 Write chemical equations for the reactions involved in the Mond process for purifying nickel.

3.206 What is the oxidation number of nickel in $Ni(CO)_4$? Which is the most important oxidation state of nickel?

3.207 Write a balanced net ionic equation for the reaction of nickel with hydrochloric acid.

3.208 What is the formula for the ion that gives many nickel compounds a green color in aqueous solution?

3.209 What metals are usually found in stainless steel?

3.210 What color is glass to which nickel compounds have been added?

Chemistry of Some Important Transition Elements: The Coinage Metals, Zinc, Cadmium, and Mercury

3.211 Which metals are the coinage metals?

3.212 What are some typical uses for copper?

3.213 What copper compound is found in copper ores?

How is it treated to recover the metal? Describe how impure copper is purified. (If necessary, refer back to Chapter 18.)

3.214 How does the electrical conductivity of silver compare to that of other metals?

3.215 What metal is alloyed with silver in sterling silver and jewelry silver?

3.216 What is gold leaf? What is the thickness in millimeters and in inches of one layer of gold leaf?

3.217 What oxidation states are observed for copper, silver, and gold?

3.218 Write the electron configurations of Cu, Ag, and Au.

3.219 Write electron configurations for the common oxidation states of copper, silver, and gold.

3.220 Complete and balance the following equations. If no reaction occurs, write N.R.
(a) $Cu + H_2SO_4$ (dilute) $\rightarrow$
(b) $Cu + H_2SO_4$ (hot, concd) $\rightarrow$
(c) $Cu + HNO_3$ (dilute) $\rightarrow$
(d) $Cu + HNO_3$ (concd) $\rightarrow$

3.221 What is the formula of the copper compound that gives old copper objects a green color?

3.222 How can CuCl be prepared? Write a chemical equation.

3.223 Why are copper sulfate crystals blue?

3.224 Why are concentrated solutions of $CuCl_2$ green? What happens to the color when the solution is diluted? Write chemical equilibria for the formation of the complexes involved and explain why the color change occurs based on Le Châtelier's principle.

3.225 Write chemical equations that show what happens when concentrated NaOH is added gradually to a solution of copper sulfate.

3.226 Based on what you have learned in Section 3.7, predict the formula of the complex that is formed by Cu^{2+} and CN^-.

3.227 What test can be used to detect the presence of copper ion in a solution? What are the formulas of the complex ions of copper involved in this test?

3.228 Write chemical equations for the reaction of silver with concentrated and dilute nitric acid.

3.229 What compound is formed when NaOH is added to a solution of silver nitrate? Write its formula.

3.230 How do the solubilities of the silver halides vary?

3.231 Silver bromide can be made by stirring a suspension of solid AgCl in a NaBr solution. Write the equilibria involved and explain how this reaction occurs.

3.232 Write the chemical equation involved in the analytical test for Ag^+.

3.233 Gold(III) is a fairly potent oxidizing agent. Why?

3.234 Explain, in terms of the lanthanide contraction, why gold is a much less reactive element than either copper or silver.

3.235 What are the ores of zinc, cadmium, and mercury?

3.236 Even though zinc is an active metal, it corrodes very slowly. What is the formula of the compound responsible for this?

3.237 Write net ionic equations that show what happens when zinc dissolves in HCl and NaOH.

3.238 Write net ionic equations that show what happens when concentrated base is added gradually to a solution of $ZnCl_2$.

3.239 How does a coating of zinc protect iron from corrosion, even if the coating is partially worn away? Why doesn't tin afford the same protection? (Hint: If necessary, refer to Chapter 18.)

3.240 Under what conditions would cadmium protect steel from corrosion but zinc would not?

3.241 Zinc is such an active metal that when it reacts with dilute nitric acid, the reduction product is NH_4^+. Write a balanced net ionic equation for this reaction.

3.242 Write net ionic equations for the reaction, if any, of hydrochloric acid with (a) zinc, (b) cadmium, and (c) mercury.

3.243 Write the net ionic equation for the reaction of mercury with concentrated nitric acid.

3.244 What is an *amalgam*? How is mercury used in the separation of gold from its ores?

3.245 Hg_2Cl_2 is much less toxic than $HgCl_2$ if taken internally. Why?

3.246 What are the structures of Hg_2Cl_2 and $HgCl_2$?

3.247 What reaction occurs when aqueous ammonia is added to Hg_2Cl_2? What is the color of the solid produced in this reaction?

3.248 What is *vermilion*?

Appendix A

Answers to Selected Exercises

CHAPTER 1

Review Exercises

1.9 (a) $NaH(s) + H_2O \rightarrow H_2(g) + NaOH(aq)$
(b) $CaH_2(s) + 2H_2O \rightarrow 2H_2(g) + Ca(OH)_2(aq)$
(c) $HCl(g) + H_2O \rightarrow H_3O^+(aq) + Cl^-(aq)$
(d) $2Na(s) + 2H_2O \rightarrow H_2(g) + 2NaOH(aq)$
(e) $Mg(s) + H_2(g) \rightarrow MgH_2(s)$

1.11 Ionic: MgH_2, KH, CaH_2 Molecular: H_2Se, HI, PH_3

1.20 (a) alkali metals (b) alkali metals (c) Group IIIA (d) Group VIIA

1.23 Covalent. Ionic oxides are basic and react with acids.

1.26 It must have an oxygen–oxygen bond.

1.28 −1

1.31 (a) $H_2O_2 + H_2SO_3 \rightarrow H_2O + SO_4^{2-} + 2H^+$
(b) $E^\circ_{cell} = 1.60$ V (c) 1.42 g H_2O_2

1.34 (a) $2NaH(s) + O_2(g) \rightarrow Na_2O(s) + H_2O$
(b) $H^-(s) + H_2O \rightarrow H_2(g) + OH^-(aq)$
(c) $2HgO(s) \xrightarrow{heat} 2Hg(l) + O_2(g)$
(d) $2KClO_3(s) \xrightarrow{heat} 2KCl(s) + 3O_2(g)$
(e) $Na_2O_2(s) + 2H_2O \rightarrow 2NaOH(aq) + H_2O_2(aq)$
(f) $4Li(s) + O_2(g) \rightarrow 2Li_2O(s)$
(g) $2H_2O_2(l) \rightarrow 2H_2O + O_2(g)$

1.40 More difficult, because the NH_3 would vaporize more *rapidly*.

1.41 (a) $NH_3(aq) + H_2O \rightleftharpoons NH_4^+(aq) + OH^-(aq)$
(b) $NH_3(aq) + H_3O^+(aq) \rightarrow NH_4^+(aq) + H_2O$
(c) $4NH_3(aq) + 3O_2(g) \rightarrow 2N_2(g) + 6H_2O$

1.42 (a) $NH_3(aq) + HCl(aq) \rightarrow NH_4Cl(aq)$
(b) $NH_3(aq) + HBr(aq) \rightarrow NH_4Br(aq)$
(c) $NH_3(aq) + HI(aq) \rightarrow NH_4I(aq)$
(d) $2NH_3(aq) + H_2SO_4(aq) \rightarrow (NH_4)_2SO_4(aq)$
(e) $NH_3(aq) + HNO_3(aq) \rightarrow NH_4NO_3(aq)$

1.47 $KNH_2(s) + H_2O \rightarrow KOH(aq) + NH_3(aq)$
$NH_2^-(s) + H_2O \rightarrow OH^-(aq) + NH_3(aq)$

1.49 $NH_4^+(aq) + H_2O \rightleftharpoons NH_3(aq) + H_3O^+(aq)$

1.52 (a) $Mg_3N_2(s) + 6H_2O \rightarrow 2NH_3(g) + 3Mg(OH)_2(s)$
(b) magnesium oxide; $2Mg(s) + O_2(g) \rightarrow 2MgO(s)$

1.55 Poisonous hydrazine can be formed.

1.58

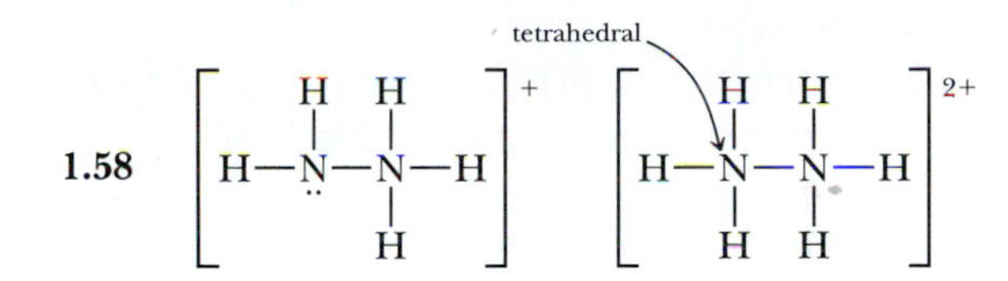

1.60 ammonia

1.63 basic; $N_3^-(aq) + H_2O \rightleftharpoons HN_3(aq) + OH^-(aq)$

1.73 Decomposition into N_2 and O_2 supplies expanding gases.

1.74 NO_2 has an unpaired electron on N; N_2O_4 has no unpaired electrons.

1.76 (a) 4.5×10^{-4} (b) $HNO_2(aq) + H_2O \rightleftharpoons H_3O^+(aq) + NO_2^-(aq)$

1.77 Raise the pH; $NO_2^-(aq) + H_2O \rightleftharpoons HNO_2(aq) + OH^-(aq)$

1.95 $CN^-(aq) + H_2O \rightleftharpoons HCN(aq) + OH^-(aq)$

1.104 (a) $N_2(g) + 2H_2(g) \rightarrow N_2H_4(l)$, $\Delta H^\circ_f = +50.6$ kJ mol^{-1}
(b) Thermodynamically unstable, because ΔH°_f is positive and ΔS°_f would be negative.
(c) Its decomposition is so slow that the compound is able to exist.
(d) $\Delta S^\circ_f = -331$ J mol^{-1} K^{-1}. ΔS°_f is negative because molecules are becoming more complex (of lower entropy).

1.106 (a) $NO_3^-(aq) + 2H^+(aq) + e^- \rightarrow NO_2(g) + H_2O$ (reduction)
$NO(g) + H_2O \rightarrow NO_2(g) + 2H^+(aq) + 2e^-$ (oxidation)
(b) $2NO_3^-(aq) + NO(g) + 2H^+(aq) \rightarrow 3NO_2(g) + H_2O$
(c) $E^\circ_{cell} = -0.04$ V, not spontaneous. The negative value of E°_{cell} corresponds to a positive

value of $\Delta G°$ and so to a nonspontaneous event.

(d) $E_{cell} = E°_{cell} - \frac{0.0592}{2} \log \frac{P^2_{NO_2}}{[NO_3^-]^2[H^+]^2 P_{NO}}$; as $[NO_3^-]$ becomes larger, the second term on the right becomes smaller, so E_{cell} becomes, overall, less negative. This makes $\Delta G°$ less positive and so tends to favor the forward reaction.

1.108 Net ionic equations are as follows:

(a) $H^+(aq) + OH^-(aq) \rightarrow H_2O$
(b) $H^+(aq) + HCO_3^-(aq) \rightarrow H_2O + CO_2(g)$
(c) $2H^+(aq) + CO_3^{2-}(aq) \rightarrow H_2O + CO_2(g)$
(d) same as (a)
(e) same as (c)
(f) same as (b)
(g) $CaCO_3(s) + 2H^+(aq) \rightarrow Ca^{2+}(aq) + H_2O + CO_2(g)$
(h) $Ca(OH)_2(s) + 2H^+(aq) \rightarrow Ca^{2+}(aq) + 2H_2O$
(i) $Mg(OH)_2(s) + 2H^+(aq) \rightarrow Mg^{2+}(aq) + 2H_2O$
(j) $MgCO_3(s) + 2H^+(aq) \rightarrow Mg^{2+}(aq) + H_2O + CO_2(g)$
(k) $S^{2-}(aq) + 2H^+(aq) \rightarrow H_2S(g)$
(l) $SO_3^{2-}(aq) + 2H^+(aq) \rightarrow SO_2(g) + H_2O$
(m) no reaction
(n) no reaction
(o) $CN^-(aq) + H^+(aq) \rightarrow HCN(g)$
(p) $Pb^{2+}(aq) + 2Cl^-(aq) \rightarrow PbCl_2(s)$
(q) $Ag^+(aq) + Cl^-(aq) \rightarrow AgCl(s)$
(r) $Ca(s) + 2H^+(aq) \rightarrow Ca^{2+}(aq) + H_2(g)$
(s) $NO_2^-(aq) + H^+(aq) \rightarrow HNO_2(aq)$
(t) no reaction
(u) $Mg(s) + 2H^+(aq) \rightarrow Mg^{2+}(aq) + H_2(g)$
(v) $C_2H_3O_2^-(aq) + H^+(aq) \rightarrow HC_2H_3O_2(aq)$
(w) $NaNH_2(s) + 2H^+(aq) \rightarrow Na^+(aq) + NH_4^+(aq)$
(x) $N_3^-(s) + H^+(aq) \rightarrow HN_3(aq)$
(y) $NH_3(aq) + H^+(aq) \rightarrow NH_4^+(aq)$
(z) no reaction

CHAPTER 2

Review Exercises

2.10 Sulfurous acid; sulfur dioxide, $SO_2 \cdot nH_2O$, bisulfite ion, $HSO_3^-(aq)$, and sulfite ion, $SO_3^{2-}(aq)$.

2.11 (a) $NaOH(aq) + SO_2(g) \rightarrow NaHSO_3(aq)$
(b) $NaHCO_3(aq) + SO_2(g) \rightarrow NaHSO_3(aq) + CO_2(g)$

2.16 (a) $NaHSO_3(aq) + HCl(aq) \rightarrow NaCl(aq) + H_2O + SO_2(g)$
$HSO_3^-(aq) + H^+(aq) \rightarrow H_2O + SO_2(g)$
(b) $Na_2SO_3(aq) + 2HCl(aq) \rightarrow 2NaCl(aq) + H_2O + SO_2(g)$
$SO_3^{2-}(aq) + 2H^+(aq) \rightarrow H_2O + SO_2(g)$

2.18 (a) $SO_3(g) + H_2O \rightarrow H_2SO_4(aq)$
(b) $SO_3(g) + NaHCO_3(aq) \rightarrow NaHSO_4(aq) + CO_2(g)$
$SO_3(g) + 2NaHCO_3(aq) \rightarrow Na_2SO_4(aq) + H_2O + 2CO_2(g)$
(c) $SO_3(g) + Na_2CO_3(aq) \rightarrow Na_2SO_4(aq) + CO_2(g)$
(d) $SO_3(g) + NaOH(aq) \rightarrow NaHSO_4(aq)$
$SO_3(g) + 2NaOH(aq) \rightarrow Na_2SO_4(aq) + H_2O$

2.23 CO_2, $NaHSO_4$, $2NaHSO_4(aq) + Na_2CO_3(s) \rightarrow 2Na_2SO_4(aq) + CO_2(g) + H_2O$

2.30 (a) $S_2O_3^{2-}(aq) + 4Cl_2(g) + 5H_2O \rightarrow 2HSO_4^-(aq) + 8H^+(aq) + 8Cl^-(aq)$
(b) $2S_2O_3^{2-}(aq) + I_2(aq) \rightarrow 2I^-(aq) + S_4O_6^{2-}(aq)$

2.34 pH = 6.96

2.40 $H_2SeO_4(aq) + 2NaOH(aq) \rightarrow Na_2SeO_4(aq) + 2H_2O$

2.48 $PBr_3 + 3H_2O \rightarrow H_3PO_3 + 3HBr$
$PBr_5 + 4H_2O \rightarrow H_3PO_4 + 5HBr$

2.57 (a) arsenic acid (b) sodium arsenate (c) arsenous acid (d) sodium dihydrogen arsenate (e) arsenic trichloride (f) antimony pentachloride

2.60 $2F_2 + 2H_2O \rightarrow 4HF + O_2$; $Cl_2 + H_2O \rightarrow HCl + HOCl$; $Br_2 + H_2O \rightarrow HBr + HOBr$; $I_2 + H_2O \rightarrow HI + HOI$

2.68 (a) *Y* (actually, Y^-) (b) *X* (actually, X_2) (c) No, Br_2 is a weaker oxidizing agent than Cl_2. (d) Yes, Cl_2 is a stronger oxidizing agent than Br_2. (e) Yes, Br_2 is a stronger oxidizing agent than I_2.

2.79 (a) P_2S_5 (b) P_4S_{10}

2.80 (a) $IO_3^- + 3HSO_3^- \rightarrow I^- + 3SO_4^{2-} + 3H^+$
(b) $5I^- + IO_3^- + 6H^+ \rightarrow 3I_2 + 3H_2O$
(c) 3.79 g $NaHSO_3$ (d) 7.98 g brine

2.81 $K_{eq} = \frac{[H^+][Cl^-][HOCl]}{[Cl_2]} = 0.015$

CHAPTER 3

Review Exercises

3.3 (a) Tl (b) Ba (c) K

3.5 MgO

3.9 It is not economically feasible to extract aluminum from aluminosilicates.

3.43 (a) $2Li + 2H_2O \rightarrow 2LiOH + H_2(g)$

(b) $2K + 2H_2O \rightarrow 2KOH + H_2(g)$
(c) $2Rb + 2H_2O \rightarrow 2RbOH + H_2(g)$

3.45 (a) $Na_2O_2 + 2H_2O \rightarrow 2NaOH + H_2O_2$
(b) $Na_2O + H_2O \rightarrow 2NaOH$
(c) $2KO_2 + 2H_2O \rightarrow 2KOH + O_2 + H_2O_2$

3.53 $CO_3^{2-} + H_2O \rightleftharpoons HCO_3^- + OH^-$, pH = 12.16

3.54 (a) $NaClO_4$ (b) $KC_2H_3O_2$ (c) $KC_2H_3O_2$

3.57 1.41×10^5 lb

3.77 $Mg(OH)_2$. $Mg(OH)_2 + 2HCl \rightarrow MgCl_2 + 2H_2O$

3.80 $2Mg + O_2 \rightarrow 2MgO$; $MgCO_3 \xrightarrow{\text{heat}} MgO + CO_2$;
$Mg(OH)_2 \xrightarrow{\text{heat}} MgO + H_2O$

3.116 PbO_2

3.121 $Pb(OH)_4^{2-} + OCl^- \rightarrow PbO_2(s) + Cl^- + 2OH^- + H_2O$

3.122 $PbO_2(s) + 4Cl^-(aq) + 4H^+(aq) \rightarrow PbCl_2(s) + Cl_2(g) + 2H_2O$

3.126 (a) 1.6×10^{-2} *M* (b) 8.1×10^{-3} *M* (c) 1.3×10^{-3} *M*

3.128 The halide ion is oxidized by Pb^{IV}.
$PbBr_4 \rightarrow PbBr_2 + Br_2$

3.142 16.7 g/cm^3

3.147 Magnetic domains within the solid shift to align with Earth's magnetic field.

3.154 $Bi^{3+} + H_2O \rightarrow [Bi{-}OH_2]^{3+} \rightarrow BiO^+ + 2H^+$

3.162 $Cr(H_2O)_6^{3+} + H_2O \rightleftharpoons Cr(H_2O)_5OH^{2+} + H_3O^+$

3.166 $H_2CrO_4 \rightleftharpoons H^+ + HCrO_4^-$; $HCrO_4^- \rightleftharpoons H^+ + CrO_4^{2-}$

3.173 (a) $Mn(s) + 2HCl(aq) \rightarrow MnCl_2(aq) + H_2(g)$
(b) $MnCl_2(aq) + 2NaOH(aq) \rightarrow Mn(OH)_2(s) + 2NaCl(aq)$

3.187 $Fe(s) + 2H^+(aq) \rightarrow Fe^{2+}(aq) + H_2(g)$

3.193 $Cr(H_2O)_3(OH)_3$ is able to dissolve in base, but $Fe(H_2O)_3(OH)_3$ is not.

3.194 The Fe^{3+} ion has more of a polarizing effect on a neighboring water molecule than does the larger and less highly charged Fe^{2+} ion.

3.198 Add Fe^{2+} to the solution suspected to contain CN^-. If CN^- is present it will form $Fe(CN)_6^{4-}$. Then add Fe^{3+} ion. A blue precipitate of Prussian blue confirms the presence of CN^-.

3.203 $4Co^{3+}(aq) + 2H_2O \rightarrow 4Co^{2+}(aq) + O_2(g) + 4H^+(aq)$

3.216 Very thin gold foil; 9.09×10^{-5} mm, 3.58×10^{-6} in.

3.220 (a) N.R.
(b) $Cu + 2H_2SO_4 \rightarrow CuSO_4 + SO_2 + 2H_2O$
(c) $3Cu + 8HNO_3 \rightarrow 3Cu(NO_3)_2 + 2NO + 4H_2O$
(d) $Cu + 4HNO_4 \rightarrow Cu(NO_3)_2 + 2NO_2 + 2H_2O$

3.225 $Cu^{2+} + 2OH^- \rightarrow Cu(OH)_2(s)$; $Cu(OH)_2(s) + 2OH^- \rightarrow Cu(OH)_4^{2-}$

3.226 $Cu(CN)_4^{2-}$

3.228 Concentrated HNO_3: $Ag + 2HNO_3 \rightarrow AgNO_3 + NO_2 + H_2O$; dilute HNO_3: $3Ag + 4HNO_3 \rightarrow 3AgNO_3 + NO + 2H_2O$

3.233 Because gold is so easily reduced.

3.238 $Zn^{2+}(aq) + 2OH^-(aq) \rightarrow Zn(OH)_2(s)$
$Zn(OH)_2(s) + 2OH^-(aq) \rightarrow Zn(OH)_4^{2-}(aq)$

3.241 $4Zn(s) + NO_3^-(aq) + 10H^+(aq) \rightarrow 4Zn^{2+}(aq) + NH_4^+(aq) + 3H_2O$

3.243 $Hg(l) + 2NO_3^-(aq) + 4H^+(aq) \rightarrow Hg^{2+}(aq) + 2NO_2(g) + 2H_2O$

Photo Credits

CHAPTER 1

Opener: Tass/Sovfoto/Eastfoto. *Figure 1.1:* Michael Watson. *Page 4:* OPC, Inc. *Page 5:* Courtesy NASA. *Figure 1.2a:* OPC, Inc. *Figure 1.3:* Courtesy NASA. *Figure 1.4:* Ray Pfortner/Peter Arnold, Inc. *Figure 1.5:* Michael Watson, *Page 15:* Robert Capece. *Figure 1.6:* OPC, Inc. *Figure 1.7:* Thomas Horland/ Grant Heilman Photography. *Figure 1.8:* OPC, Inc. *Figure 1.9:* Peter Lerman. *Page 21 (top):* Michael Watson. *Figure 1.10:* Ken Karp/OPC, Inc. *Figure 1.12:* Jim Mendenhall. *Figure 1.13:* Andy Washnik. *Page 30:* Courtesy Smithsonian Institution. *Page 22:* Ken Karp. *Page 24 (top):* OPC, Inc. *Page 24 (bottom):* Courtesy Ireco. *Page 34 (top):* Astrid & Hanns-Frieder Micheler/ Science Photo Library/Photo Researchers. *Page 34 (bottom):* Mula & Haramaty/Phototake.

CHAPTER 2

Opener: Christian Crzimek/Okapia/ Photo Researchers. *Figure 2.1:* Yale Peabody Museum. *Figure 2.3:* OPC, Inc. *Figure 2.4: Ken Karp. Figure 2.5:* Michael Watson. *Pages 47 and 49 (top):* Andy Washnik. *Figure 2.6:* OPC, Inc. *Figures 2.7 and 2.8:* Andy Washnik. *Figure 2.10:* Wesley Frank/Woodfin Camp & Associates. *Figures 2.11 and 2.12:* Michael Watson. *Figure 2.13:* OPC, Inc. *Figures 2.14 and 2.15:* Michael Watson. *Figure 2.16:* Ginger Chih/Peter Arnold, Inc. *Page 59:* Courtesy of Dr. Noble R. Usherwood/Potash & Phosphate Institute. *Page 61 (top center):* Yale Peabody Museum. *Page 61 (top):* Michael Watson. *Page 61 (bottom center):* Yale Peabody Museum. *Page 62 (bottom):* Brian J. Skinner. *Page 64 (bottom):* OPC, Inc. *Page 78 (bottom):* Robert Capece.

CHAPTER 3

Opener: John Lund/Tony Stone Images/New York, Inc. *Figure 3.1:* Courtesy Smithsonian Institution. *Figure 3.2:* Courtesy NOAA. *Page 82:* Peter B. Kaplan/Photo Researchers. *Figure 3.4:* Courtesy NASA. *Page 84:* David M. Campione/Science Photo Library/ Photo Researchers. *Figure 3.5:* Courtesy Bethlehem Steel. *Page 86:* J. M. Mejuto. *Page 88 (left):* Courtesy AKZO Salt Inc. *Page 88 (right):* Tom Till. *Figure 3.8:* Yoav/Phototake. *Page 92:* OPC, Inc. *Figure 3.9 (left):* Fred Ward/Black Star. *Figure 3.9 (right):* Al Hamdan/The Image Bank. *Figures 3.10 and 3.11:* OPC, Inc. *Page 96 (top right):* Courtesy Dow Chemical. *Page 98:* Michael Watson. *Figure 3.14:* Science Photo Library/Photo Researchers. *Figure 3.15:* Yoav/Phototake. *Figure 3.16:* Andy Washnik. *Figure 3.17:* Courtesy Orgo-Thermit, Inc. *Figure 3.18:* Courtesy NASA. *Figure 3.19:* Andy Washnik. *Figure 3.21:* Michael Watson. *Figure 3.22:* Fred Ward/Black Star. *Figure 3.23:* Michael Watson. *Figure 3.25:* OPC, Inc. *Figure 3.26:* OPC, Inc. *Figure 3.30:* Michael Watson. *Figure 3.31:* OPC, Inc. *Figures 3.33, 3.34, 3.35, 3.37 and 3.38:* Michael Watson. *Page 119 (bottom):* Michael Watson. *Figure 3.39:* Peter Hendrie/The Image Bank. *Figure 3.40:* Andy Washnik. *Figure 3.41:* OPC, Inc. *Figure 3.42:* Michael Watson. *Figure 3.42:* Courtesy Safra Nimrod. *Figure 3.43 and 3.44:* Michael Watson. *Page 124 (top):* OPC, Inc. *Page 125:* H. W. Kitchen, National Audubon Society. *Page 127:* Courtesy USDA. *Page 129:* William Sacco/Yale Peabody Museum. *Figure 3.45:* Wesley Bocxe/ Photo Researchers. *Figure 3.46:* Andy Washnik.

Index

Numbers set in italics refer to tables.

D

R

S

X

Z